Stochastic Models of Systems

Mathematics and Its Applications

Managing Editor:

M. HAZEWINKEL

Centre for Mathematics and Computer Science, Amsterdam, The Netherlands

Volume 469

Stochastic Models of Systems

by

Vladimir S. Korolyuk

and

Vladimir V. Korolyuk

Institute of Mathematics,
Ukrainian Academy of Sciences,
Kiev, Ukraine

SPRINGER-SCIENCE+BUSINESS MEDIA, B.V.

A C.I.P. Catalogue record for this book is available from the Library of Congress.

ISBN 978-0-7923-5606-6   ISBN 978-94-011-4625-8 (eBook)
DOI 10.1007/978-94-011-4625-8

Printed on acid-free paper

Contents

Preface		**ix**
1	**Introduction**	**1**
1.1	Classification and properties of stochastic systems	1
	1.1.1 Phase states space	2
	1.1.2 Evolution of stochastic systems	3
	1.1.3 Stochastic systems	4
	1.1.4 Semi-Markov and Markov systems	5
	1.1.5 Simplified description of the system	8
1.2	Renewal processes	11
	1.2.1 Renewal equation	12
	1.2.2 Auxiliary processes	13
	1.2.3 Transformation of recurrent flows	15
	1.2.4 Superposition of renewal processes	16
	1.2.5 The alternating renewal process	19
2	**Markov renewal processes**	**21**
2.1	Definition of Markov renewal process	22
	2.1.1 Homogeneous Markov chain	23
	2.1.2 Semi-Markov kernel	24
	2.1.3 Markov Renewal Process	24
	2.1.4 Markov jump processes	26
	2.1.5 Stochastic representation of MRP	28
2.2	Semi-Markov processes	30

      2.2.1   Markov renewal equation 31

      2.2.2   Characteristics of MRP 32

      2.2.3   Auxiliary processes 34

      2.2.4   Regenerative processes 35

  2.3   Ergodicity and stationary
distribution 36

      2.3.1   Ergodic property of Markov chain 37

      2.3.2   Ergodicity of semi-Markov process 38

3 Phase merging algorithms   **43**

  3.1   Reducible-invertible operators 43

  3.2   Perturbation of reducible-invertible
operators 48

  3.3   Martingale characterization
of Markov processes 56

  3.4   Pattern limit theorem 62

  3.5   Ergodic phase merging 66

  3.6   Splitting phase merging 71

  3.7   Heuristic phase merging principles 76

      3.7.1   Basic assumptions 77

      3.7.2   Heuristic phase merging principles 78

      3.7.3   Heuristic phase merging 81

**4 Evolutional stochastic system
in a random medium**   **89**

  4.1   Stochastic additive functionals 89

  4.2   Storage Processes 96

  4.3   Random evolution 100

  4.4   Ergodic average and diffusion approximation of random evolutions 105

  4.5   Splitting average and diffusion approximation of random evolutio n 109

  4.6   Application of average and diffusion approximation algorithms 114

  4.7   Counting processes 121

  4.8   Proofs of limit theorems 127

5 Diffusion approximation of Markov queueing systems and networks **137**

5.1 Algorithms of diffusion approximation 137

5.2 Markov queueing processes 139

5.3 Average and diffusion approximation 144

    5.3.1 Average scheme 145

    5.3.2 Diffusion approximation scheme 147

5.4 Stationary distribution 152

5.5 Markovian queueing systems 157

    5.5.1 Collective limit theorem in R^1 157

    5.5.2 Systems of M/M type 160

    5.5.3 Systems with bounded input 161

    5.5.4 Repairman problem 163

5.6 Markovian queueing networks 170

    5.6.1 Collective limit theorems in R^N 170

    5.6.2 Markov queueing networks 175

References **181**

Preface

The notion of the system is so uncertain that a general theory of systems, which would cover various schemes and methods of analysis, could not be constructed.

Natural classification of mathematical models of systems appears due to separation of the sufficiently general and, at the same time, common methods of analysis.

Among the boundless variety of stochastic models of systems whose evolution develops under the influence of random factors have been selected. Moreover, it has been assumed that the evolution takes place in a *random medium,* i.e. *unilateral interaction between the system and the medium.* Local characteristics of the system are changing with the change of the state of the random medium but the evolution of the system does not affect the stochastic conformity of the medium.

This particular feature of the interaction between a system and a medium essentially narrows down the totality of actual systems which possess the above mentioned property. However, the unified effective mathematical methods for analysis remain valid.

At the same time, in order to avoid additional, sometimes essential, complications in the analysis, only *Markovian models of the random medium* are considered in this book although the analogous results of analysis of evolution of the stochastic system take place in a *semi-Markov random medium* as well.

Thus stochastic models of systems are determined by two processes: a *switched process* describing the evolution of the system and a *switching process* describing changes of the random medium.

It is assumed that the mathematical model of evolution of the system has a semi-group property and the random medium has an ergodic property.

Enumerated properties of stochastic models of systems extract a class of systems representing *Markov random evolutions.*

The construction of an adequate mathematical model to real complicated systems leads to a knotty model whose analysis is difficult to perform even with the help of modern computers. The "tyranny of dimensionality" of a mathematical model is special characteristic of stochastic evolution systems in a random medium. Potentially, unbounded fluctuation of random factors of the medium adds to the complexity of the mathematical model.

In this situation the *problem of simplified description of evolution of the system* arises. The most effective and natural way of analysis of stochastic models of systems arises from a locally adequate mathematical model, but, in this case, it is necessary to apply mathematically justified methods of simplification of the initial model.

It is worthy to repeat the words of the well-known specialist on systems theory Walter Ethbey (1969): "... theory of systems should be built on the methods of simplification and is, essentially, the science of simplification."

In the theory of random evolutions, the limit average theorems in series schemes can be regarded as the effective tools of the simplified description of evolutional systems in a random medium.

In this connection, the ergodic properties of the random medium and semi-group property of evolution of the system are essential.

Effective mathematical tools of analysis are based on the theory of singular perturbed reducibly invertible operators and on the martingale characterization of Markov random processes.

Limit average theorems are based on the weak convergence of Markov processes.

The peculiarity of the consideration of random evolutions is the supposition that the Markov process which describes a random medium is considered *in series schemes* as well. Such a supposition gives a new type of *phase merging algorithms.*

Stochastic models of systems analysis are discussed in this book in all aspects and at all stages, beginning with the construction of mathematical models of real systems, including mathematical analysis of models based on the simplification methods and finishing with interpretation of the results in real stochastic systems.

In Chapter 1 (Introduction) the classification and basic properties of stochastic systems are considered using a rather elementary approach. The simplification of the stochastic systems is discussed as a phase merging principles. The phase merging scheme consists in the splitting of the phase space of states and the merging of the separate classes of states in the individual merged states of the merged system.

In Chapter 2 (Markov renewal processes) the stationary jump Markov and semi-Markov processes are considered being defined by Markov renewal processes. The choice of Markov and semi-Markov jump processes as a mathematical models of stochastic systems is determined by the constructive possibilities of the effective definition and analysis by means of the martingale characterization (Section 3.3). The phase merging algorithms are considered in Chapter 3. The verification of the phase merging algorithms is based on the asymptotical solution of the singular perturbation problems for the reducible-invertible operators (Section 3.1 and 3.2). In addition, the heuristic principles of phase merging for some classes of systems which can be useful in applications are considered (Section 3.7). The heuristic phase merging of the stochastic system considered in Section 3.7.3, gives the same results as the phase merging algorithms based on the limit theorem.

In Chapter 4 the evolutional stochastic systems in a random medium are described by the random evolutions as the operator-valued random processes. The storage process and the stochastic additive functionals of the Markov jump processes are interpreted as the random evolutions (Section 4.1, 4.2).

The ergodic and splitting average algorithms, as well as diffusion approximation, are considered in Sections 4.4, 4.5.

The final Chapter 5 is devoted to the actual problem of diffusion approximation of stochastic systems, which can be described the Markov processes with locally independent increments taking values in the Euclidean vector space. The applications of pattern and collective limit theorems to Markov queueing systems and networks are considered.

The average and diffusion approximation algorithms for evolution systems in a random medium discussed, naturally do not possess exhausting generality.

The tendency to present accessible methods of analysis of random evolutions and save essential analytical problems appearing in this analysis makes authors to confine themselves to the analysis of Markov random evolutions only. That is why this book can be regarded as an introduction to the theory of random evolutions guided to applied aspects of this theory. The book will be useful to specialists in system analysis who apply mathematical models of evolutional systems under the influence of random factors in their practice.

Chapter 1

Introduction

Among the variety of models of real systems we can distinguish those whose evolution develops under the influence of random factors. The main properties of stochastic systems and concepts (principles) of phase merging which simplify description and analysis of systems, are discussed.

1.1 Classification and properties of stochastic systems

In a mathematical description of real systems it is necessary to determine the basic quantitative system parameters (characteristics) which completely determine the behaviour of the system under given conditions.

The choice of the parameters of a system will be considered as basic and is a very important starting problem in the construction of mathematical models.

First of all, we should take into consideration the possibility of observing the modification of parameters of the system as well as the final goal of the analysis of the mathematical model to obtain objective quantitative characteristics which determine the essential properties of the evolution of the system.

Depending on the conditions of observation and taking into consideration quantitative changes and the final goal, the choice of basic parameters which characterize the behaviour of one and the same real system can be different, more or less detailed and merged.

When constructing a mathematical model of the system, we will suppose that this choice has already been made, and what we need to do is to describe the set of all possible values of parameters of the system.

1.1.1   Phase states space

A fixed value of a basic parameter is determined by the system's state. The set of all possible different states of the system is called the *phase states space*.

A mathematical model of a real system contains the phase space as starting object in the description. We can take any set of elements of arbitrary nature as the phase space.

For example, an alphabet (in letters, numeric or symbolic). The set of integer numbers (finite or infinite) is used as a phase space for discrete systems.

The set of real numbers and set of vectors (finite collection of numbers) are used as a phase space for systems with continuous set of possible states.

In short, every set having enough elements to describe (to code) different states of the system to be measured or observed can be used as a phase space.

Additional phase states, which play an important role in the mathematical description of the system, among physical (observable) states, may be useful in the construction of a mathematical model.

For example, we can not only use space coordinates as phase coordinates (states) of material particles, but the velocities of particles as well. Additional space coordinates are necessary for uniqueness of description of the system's evolution in time. In mathematical models of stochastic systems additional phase states are necessary in order to apply mathematical methods developed in the theory of Markov random processes.

The most general form of the phase space of a system is given by a measurable space $(E, \mathcal{E})$, where E is the set of all states of the system (finite, countable or uncountable set) and $\mathcal{E}$ is Boolean σ-algebra of measurable subsets of E whose elements represent observable sets of states of the system.

In particular, we suppose that separate states $x \in E$ are observable, i. e. that single-point sets belong to σ-algebra $\mathcal{E}$: $\{x\} \in \mathcal{E}$. As it usually occurs the Boolean σ-algebra $\mathcal{E}$ has the following property: for every sequence of sets $\Delta_n \in \mathcal{E}, n \geq 1$, the sum $\cup_{n \geq 1} \Delta_n \in \mathcal{E}$ and intersection $\cap_{n \geq 1} \Delta_n \in \mathcal{E}$ and, besides that, along with $B \in \mathcal{E}$ the complement set $\bar{B} := E \backslash B \in \mathcal{E}$. In particular, $E \in \mathcal{E}$ implies that the empty set $\emptyset \in \mathcal{E}$.

Often the mathematical analysis of a stochastic system is simplified by the assumption that σ-algebra $\mathcal{E}$ is *countably generated*. It means that we can select countable sequence of elements $B_n \in \mathcal{E}$, $n \geq 1$, such that an arbitrary element $B \in \mathcal{E}$ is represented as the sum $B = \cup_{k \geq 1} B_{n_k}$ or as the product $B = \cap_{k \geq 1} B_{n_k}$ of countable number of selected subsets.

In stochastic models of systems under consideration we mainly use the finite-dimensional Euclidean space R^N, where N is dimension of the space, as phase states space, or subset of the Euclidean space. In Euclidean space σ-algebra of Borel subsets is countably generated by parallelepipeds with rational coordinates of vertexes.

1.1.2 Evolution of stochastic systems

The modification of states of the system in time determines the evolution (functioning) of the system.

There are two fundamentally different types of evolution: *continuity* and *discrete* (impulsiveness).

Naturally these two types of evolution can be combined in one system.

Spasmodical modification of the states means that the system occupies every possible state for a finite time and then jumps into another state.

Of course, the assumption of spasmodical modification of the state is limited, but not so much as it may appear at first sight.

Actually, in real systems the transition from one observable state into another takes some time. If in this case intermediate states are not observable or are not used in analysis then the time of transition can be included into state sojourn time. Then we can treat the transition as immediate.

Besides that, the situations when the time of states transition is small enough and its discount is not essential for analysis are quite frequent.

In this case the transition from one state into another can be considered as immediate.

Continuous states modification in system means that there exists finite velocity of states transition. In this case the set of states of the system forms, as a rule, a metric space, for example, Euclidean space.

There exists an important class of stochastic systems described by diffusion stochastic processes which have finite characteristics of evolution, namely, the drift and the diffusion coefficient. The modification of states of this system in time is completely determined by these characteristics.

1.1.3   Stochastic systems

Our basic assumption in mathematical analysis here is the stochastic character of evolution of the system.

We mean random character of state transition as well as random sojourn time in state.

Stochasticity of the system does not eliminate the possibility of deterministic modification of states but, in this case, a completely deterministic evolution of the system is not considered because essential properties of the stochastic system is developed due to the random character of evolution of the system.

Markovian systems. An essential assumption on the evolution of the stochastic system which is used in mathematical analysis is *homogeneity in time.* By this we mean that the transition probabilities from state to state and the distribution of time during which the system occupies the state does not depend on a preceding evolution of the system. Due to the homogeneity of the system in time, stochastic models of systems have constructive character and accessible mathematical tools of analysis.

Nevertheless in modern theory of stochastic systems there are sufficiently effective methods of the analysis of non-homogeneous in time evolutions of systems. Homogeneity in time of stochastic systems essentially simplifies the mathematical tools of analysis, allows to understand better the properties that develop in systems over increasing time intervals.

Therefore the principal assumption regarding the evolution of stochastic systems, i.e. Markovian transitions and states of system, means the independence of probabilities of transitions and of distributions of sojourn time from the preceding evolution of the system.

Evolution of a stochastic system is determined by *Markov renewal process* $(s_n, \theta_n; \; n \geq 0)$ of which the first component fixes the state of the system after nth transition and the second component θ_n fixes the sojourn time in the state s_{n-1}.

1.1.4 Semi-Markov and Markov systems

Semi-Markov systems are described by Markov renewal processes with arbitrary distributed sojourn times.

Markov systems are described by Markov renewal processes with exponentially distributed sojourn times.

The distinction between these two processes is essential but a constructive determination with the help of Markov renewal processes is unifying.

In the same time Markov systems have additional properties which enrich mathematical tools of analysis, which is why Markov systems are mainly considered. Hence for semi-Markov systems the methods of application of algorithms of analysis developed for the Markovian systems will be described.

Evolution systems. In the analysis of a system it is necessary to take into account certain parameters which depend on the trajectory preceding to this moment of time. The parameters of this kind we call *evolution parameters.*

We consider evolution systems of two types: *dynamic systems* and *stochastic additive functionals.* Certainly, there exist evolution systems that have dynamic and stochastic components. These processes are called *storage processes.*

Transition graph and evolution diagram. A clear idea of the evolution of the system can be demonstrated by a transition graph (Fig. 1.1). Suppose that the phase space consists of three states.

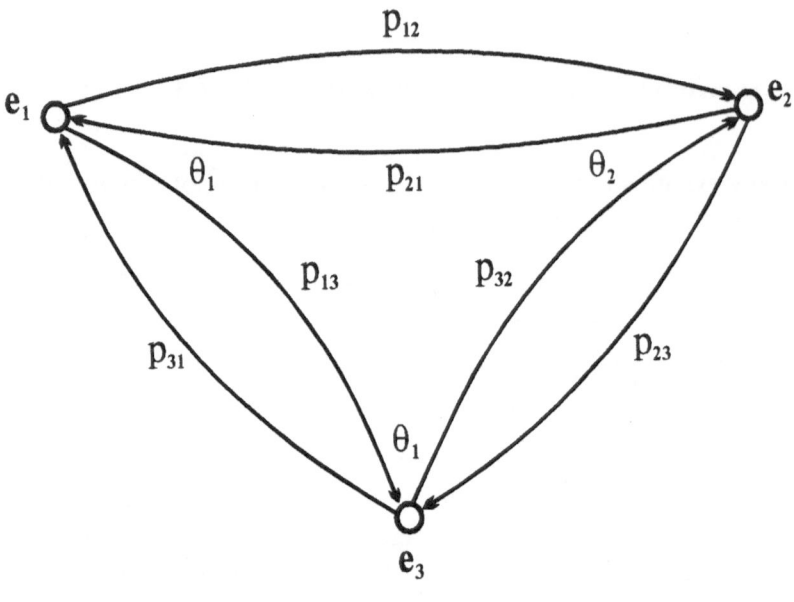

Fig. 1.1. **Transition graph of a system**

The transition between states are indicated by lines with arrows that determines oriented edges of the graph. Near every edge the transition probability p_{ij} from state i to j can be indicated. The vertices of the graph represent the states of the system. The sojourn time θ_j can be indicated near every vertex. We use different notations of the state:

e_i — in graphs and i — in text as the index of transition probability. Evolution in time is demonstrated on Fig. 1.2.

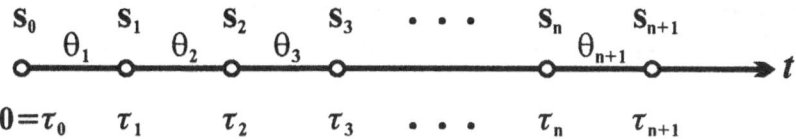

Fig. 1.2. **Evolution diagram**

We mark the moments of time when the system changes the state, by τ_n, $n \geq 0$, on real axes. The corresponding states s_n, $n \geq 0$, which are determined immediately after a change of state, can be indicated as well. The intervals between moments of change of state

$$\theta_{n+1} = \tau_{n+1} - \tau_n, \quad n \geq 0,$$

determine the *sojourn time in state* s_n.

Therefore, for the sojourn time in the state we use different notations: θ_{n+1} is the sojourn time after nth transition and θ_i is the sojourn time in state i. But it will always be clear from the text what notion is used.

In the case when the system consists of several elements (subsystems) the moments of the change of state are represented separately. (Fig. 1.3)

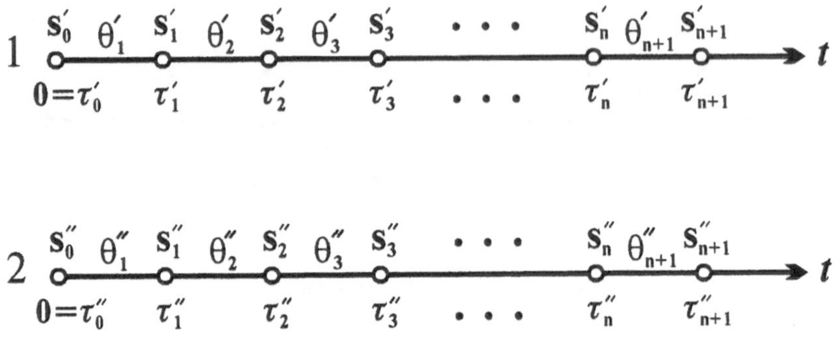

Fig. 1.3. **Evolution diagram for two-element system**

1.1.5 Simplified description of the system

In construction of a mathematical model of the system we aim for an adequate description of real systems. This causes a complication with the mathematical model. Together with this the mathematical tools are getting complicated and it is almost impossible to consider their practical usage. The principal difficulties in modelling and analysis of the system are caused by the complication of the phase space which lead to a practically boundless model. One of the actual problems of the modern theory of systems is the development of mathematically justified methods of construction for simplified models of systems whose analysis does not evoke impossible complications.

The characteristics of simplified models can be considered as corresponding characteristics of real models.

The idea of investigations of complicated systems by parts with final transition to a general system is the basis for many methods of analysis of complicated systems.

We propose the *phase merging of the states of a system algorithm* [12].

Phase mergence scheme. (Fig. 1.4) The phase merging scheme consists in the splitting of the phase space on classes that do not intersect each other:

$$E = \cup_{v \in V} E_v, \quad E_v \cap E_{v'} = \emptyset, \quad v \neq v',$$

and further merging these classes E_v, $v \in V$ to separate states $v \in V$.

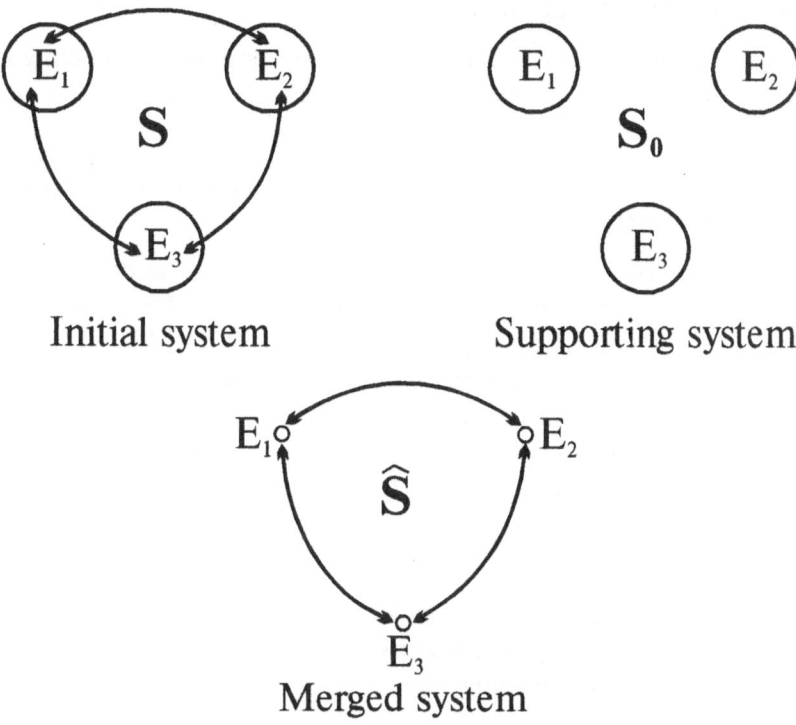

Fig. 1.4. **Phase merging scheme**

A simplified model of the system is constructed based on the merged phase space of states V.

The merged system $\widehat{S}$ is much more simple than the given system S with phase space E because only one state of the merged system corresponds to every class of states E_v of the given system.

Various connections (transitions) between states of given systems are merged to become the connection (transition) between merged states.

Therefore the analysis of the merging system is much more simplified. Moreover the principal characteristics of the simplified system can be described, with sufficient precision, such that the corresponding characteristics of the given system when splitting is successful.

It is important that the additional supporting system S^0 with the same phase space E but without connections between states is used.

To split the phase space just means introducing a new supporting system which consists of isolated subsystems S_v defined on classes of states E_v, $v \in V$.

The merged system $\widehat{S}$ is constructed with regard to ergodic properties of the supporting system S^0.

The phase merging problem, therefore, consists in the determination of the supporting system S^0 close to the given system S. Then with regard to the ergodic properties of the supporting system S^0 and local properties of the given system in the construction of a merging system $\widehat{S}$ with sufficiently precision, represents global properties of the given system S. The closeness of the given system S to the supporting system S^0 (local) signifies that connections between classes E_v are weak but not enough to be ignored because otherwise the merging brings us to a trivial result: $\bar{S}^0$ is system with isolated states $v \in V$.

The paradoxical statement: "The closer (locally) a real system is to supporting — the better (globally) a merged system describes a real system", can be explained in the following way.

The closeness of the real and supporting systems means the closeness of their local characteristics which are determined over small time intervals.

The closeness of real and merged system means the closeness of global characteristics defined for increasing time intervals.

Therefore while constructing the mathematical model of a real system we study local (differential) characteristics of the system, and in mathematical analysis we deal with global (integral) characteristics over large intervals of time. Modern natural sciences in general are based on this approach.

1.2 Renewal processes

The simplest mathematical model of real stochastic systems describe only time elapses of state for a given system. An increasing sequence of positive valued random variables $\tau_0 < \tau_1 < \ldots < \tau_n < \tau_{n+1} < \ldots$ is called a *point process* or *stochastic flow*.

The moment $\tau_n, n \geq 0$, is called a *renewal moment*. The time intervals between two adjacent renewal moments $\theta_n = \tau_n - \tau_{n-1}$, $n \geq 1$ are called *renewal lifetimes* or *sojourn times*.

The renewal process is defined by the number of renewal moments in the semi-interval $(0, t]$:

$$\nu(t) := \max \{n \geq 1 : \tau_n \leq t\}, \quad t > 0. \tag{1.1}$$

This interval contains a renewal moment only if $\tau_1 \leq t$. So that the renewal process is defined as follows:

$$\nu(t) = n, \quad \tau_n \leq t < \tau_{n+1}, \quad n \geq 0, \quad \tau_0 = 0. \tag{1.2}$$

The random process $\nu(t)$, $t \geq 0$, is also called the *counting process*.

Notice that the trajectory of the renewal process is increasing, right-continuous and is enlarging with the, equal to 1, jump.

The value of the counting process at time t is equal to the value just after the jump:

$$\nu(\tau_n) = \nu(\tau_n + 0).$$

The *point process* τ_n, $n \geq 0$, can be defined, using the counting process as follows

$$\tau(t) := \tau_{\nu(t)}, \quad t \geq 0. \tag{1.3}$$

Example 1.1. A sequence of sums,

$$\tau_n = \sum_{k=1}^{n} \theta_k, \quad n \geq 1, \tau_0 = 0,$$

made up of independent and identically distributed positive random variables θ_k, $k \geq 1$, with a common distribution function

$$F(t) = \mathcal{P} \{\theta_k \leq t\},$$

$F(0) = 0$, determines a *recurrent flow.* [5]

Example 1.2. The *Poisson flow* is given by sums of exponentially distributed renewal times:

$$\mathcal{P}\{\theta_k > t\} = e^{-\lambda t}, \quad t \geq 0.$$

The corresponding recurrent process $\nu(t)$, $t \geq 0$, is the *Poisson process* with the distribution

$$\mathcal{P}\{\nu(t) = n\} = e^{-\lambda t}(\lambda t)^n/n!, \quad n \geq 0.$$

Therefore $E\nu(t) = \lambda t$. The parameter λ determines the mean value of renewal moments when t equals 1.

At the same time λ is the *renewal intensity:*

$$\mathcal{P}\{\nu(t) = 1\} = \lambda t + o(t) \quad \text{as} \quad t \to 0.$$

1.2.1 Renewal equation

The mean value of the renewal process for the recurrent flow plays a fundamental role in the theory. It can be presented in the following form:

$$u(t) := E\nu(t) + 1. \tag{1.4}$$

This is called the *renewal function.* It can be verified that the renewal function is finite for all $t \geq 0$ with the obvious exception $F(t) \equiv 1$, $t > 0$ [5].

The renewal function is used for solving the *renewal equation*

$$\Phi(t) - \int_0^t \Phi(t-s)dF(s) = V(t), \quad t \geq 0, \tag{1.5}$$

for a given function $V(t)$, $t \geq 0$, in the following form [5]

$$\Phi(t) = \int_0^t V(t-s)du(s), \quad t \geq 0. \tag{1.6}$$

The various characteristics of Markov and semi-Markov processes and rather general regenerative processes can be presented as solutions of renewal equation (1.5).

The limit results for solution (1.6) of the renewal equation (1.5) are a principal tool and the main part of the renewal theory.

Theorem. (*Renewal*) [5][1] Let $V(t)$ be a function with a bounded variation on $R_+ = [0, +\infty)$ and a finite integral $\int_0^\infty |V(t)| dt < \infty$. Then for a non-arithmetic distribution function $F(t)$ on R_+ with a finite first moment

$$m = \int_0^\infty t dF(t) = \int_0^\infty \bar{F}(t) dt \quad (\bar{F}(t) := 1 - F(t))$$

there exists

$$\lim_{t \to \infty} \int_0^t V(t-s) du(s) = \int_0^\infty V(t) dt / m \qquad (1.7)$$

1.2.2 Auxiliary processes

For a fixed $t > 0$ let us introduce the auxiliary processes with the renewal process $\nu(t)$, $t \geq 0$:

The *remaining sojourn time (excess process)* is

$$\theta^+(t) := \tau_{\nu(t)+1} - t; \qquad (1.8)$$

The *previous sojourn time (defect process)* is

$$\theta^-(t) = t - \tau(t). \qquad (1.9)$$

The *covering sojourn time* is

$$\theta(t) := \theta^+(t) + \theta^-(t) = \theta_{\nu(t)+1}. \qquad (1.10)$$

The distribution functions of the auxiliary processes (1.8)–(1.10) satisfy the renewal equation (1.5) for certain functions $V(t)$.

Let us only consider the distribution of the remaining sojourn time

$$\Phi(t, x) := \mathcal{P}\{\theta^+(t) \leq x\} \qquad (1.11)$$

which is used in the phase merging algorithms (Chapter 2).

In order to get the right-hand side of the renewal equation (1.5) for the distribution (1.11) we have to calculate the following expectation:

---

[1] For a more general form of the renewal theorem see cf in [18]

$$
\begin{aligned}
V(t) &= EI(\theta^+(t) \le x; \tau_1 > t) \\
&= EI(\tau_1 - t \le x; \tau_1 > t) \\
&= EI(t < \tau_1 \le t + x) \\
&= F(t + x) - F(t) \\
&= \bar{F}(t) - \bar{F}(t + x).
\end{aligned}
$$

Here $I(B)$ is the indicator of event B:

$$
I(B) = \begin{cases} 1, & \omega \in B, \\ 0, & \omega \notin B. \end{cases}
$$

Therefore the distribution of the remaining sojourn time can be represented in the following way:

$$
\Phi(t, x) = \int_0^t [\bar{F}(t - s) - \bar{F}(t + x - s)] du(s).
$$

The renewal theorem yields the following limiting result:

$$
\lim_{t \to \infty} \mathcal{P}\{\theta^+(t) \le x\} = \int_0^x \bar{F}(t) dt/m. \tag{1.12}
$$

The distribution function

$$
F^*(x) := \int_0^x \bar{F}(t) dt/m =: \mathcal{P}\{\theta^* \le x\} \tag{1.13}
$$

determines the *stationary renewal time* θ^*. Notice that the distribution (1.13) has a density $f^*(t) = \bar{F}(t)/m$. The distribution (1.13) has a finite first moment only if $F(t)$ has a finite variance.

Using the stationary renewal time θ^* we can define the *stationary renewal flow* as follows:

$$
\tau_n^* = \sum_{k=0}^n \theta_k, \quad n \ge 0,
$$

$$
\tau_0^* := \theta_0 = \theta^*. \tag{1.14}
$$

It is not difficult to verify that the stationary renewal flow has a *constant renewal rate:* $u(t) = t/m$. Consequently the remaining sojourn time in the stationary renewal flow has a stationary renewal time distribution:

$$
\mathcal{P}\{\theta^+(t) \le x\} = F^*(x). \tag{1.15}
$$

This property of the renewal flow is used in the heuristic principals of phase merging (Section 3.7). The main concept is that, considering the renewal flow in 'steady-state' regime after a rather long time period T, one can conclude that the distribution of the remaining sojourn time $\theta^+(t)$ can be considered as coinciding with the distribution of the stationary renewal time θ^*.

1.2.3 Transformation of recurrent flows

The applications of recurrent flows are based on two forms of transformation: *thinning* and *superposition*. The *thinning recurrent flow* means that every renewal moment can be excluded from the flow with a fixed probability p since it is independent of a past, and remains in the flow with probability $q = 1-p$. After thinning out the initial recurrent flow τ_n, $n \geq 0$, is transformed into the *sparse flow* τ_n^*, $n \geq 0$, with renewal times

$$\theta_n^* = \tau_n^* - \tau_{n-1}^*, \quad n \geq 1.$$

It is obvious that the *thinning out flow* remains recurrent. i.e. The renewal times θ_n^*, $n \geq 1$, are independent and identically distributed. The renewal time θ_n^* can be presented in the following form:

$$\theta_1^* = \sum_{k=1}^{\nu} \theta_k \tag{1.16}$$

where ν is an integer valued random variable with a geometric distribution

$$\mathcal{P}\{\nu = n\} = qp^{n-1}, \quad n \geq 1.$$

In addition, ν and θ_k, $k \geq 1$, are mutually independent. Notice that by the Kolmogorov–Prokhorov formula [17]

$$E\theta_1^* = E\theta_k E\nu = m/q, \quad m := E\theta_k, \quad E\nu = 1/q. \tag{1.17}$$

It is easy to check that for the Poisson flow with exponentially distributed renewal times, $\mathcal{P}\{\theta_n > t\} = e^{-\lambda t}$, $n \geq 1$, the thinning out flow remains Poisson with the renewal intensity

$$\lambda^* = \lambda q. \tag{1.18}$$

Moreover, for an arbitrary recurrent flow, under sufficiently small q, the thinning out flow becomes Poisson by the Reny theorem [8]:

$$\lim_{q \to 0} \mathcal{P}\{\theta_1^* q > t\} = e^{-\lambda t}, \quad \lambda = 1/m. \tag{1.19}$$

The thinning of the recurrent flow, when applied, has important interpretations.

Let a repairing system work according to the recurrent process with the mean working time value $m = E\theta_1$ and stopping probability q being sufficiently small. Then the total working time has an exponential distribution with intensity $\lambda^* = \lambda q$ where $\lambda := 1/m$.

This principle will be used in the phase merging algorithms of the semi-Markov system (Chapter 3).

1.2.4   Superposition of renewal processes

The united renewal moments of independent renewal point processes form the superposition of renewal processes. The counting process of superposition is equal to the sum of counting processes combined in superposition:

$$\nu(t) = \sum_{k=1}^{N} \nu_k(t).$$

In the simplest case of superposition of Poisson processes the resulting process is Poisson as well with the intensity being equal to the sum of intensities of united renewal processes:

$$\lambda = \sum_{k=1}^{N} \lambda_k.$$

The superposition of renewal processes has a natural interpretation in the system theory as a mathematical model of a system with N independently functioning elements.

The heuristic principles of phase merging algorithms (Section 3.7) are applied to the procedure of superposition of renewal processes.

Here we only consider a simple example.

Example 1.3. *Superposition of two recurrent flows.* Let two independent renewal processes be given by two sequences of sums

$$\tau_n^{(i)} = \sum_{k=1}^{n} \theta_k^{(i)}, \quad n \geq 1, \quad \tau_0^{(i)} = 0, \quad i = 1, 2, \tag{1.20}$$

of independent and identically distributed positive random variables $\theta_k^{(i)}$, $k \geq 1$, $i = 1, 2$, with given distribution functions

$$P_i(t) = \mathcal{P}\{\theta_k^{(i)} \leq t\}, \quad P_i(0) = 0, \quad i = 1, 2. \tag{1.21}$$

The superposition of two recurrent flows can be given by the sum

$$\nu(t) = \nu_1(t) + \nu_2(t), \tag{1.22}$$

where $\nu_i(t) := \max\{n \geq 1 : \tau_n^{(i)} \leq t\}$, $i = 1, 2$.

It is evident that the counting process (1.22) is not defined by the recurrent flow. But a *multivariate point process* [7] which determines the counting process $\nu(t)$ can be constructed.

Note that the counting process

$$\nu(t) := \max\{n \geq 1 : \tau_n \leq t\}$$

contains two type of renewal moments belonging to different recurrent flows (1.20).

Introduce the phase space of Markov renewal times

$$E = \{ix; \ i = 1, 2; \ x > 0\} \tag{1.23}$$

where the first integer-valued component i settles the index of recurrent flow and the second continuous component settles the residual time to the adjacent renewal moment of another recurrent flow.

The state $1x$ means that the renewal moment τ_n of superposition belong to the first flow and adjacent renewal moment of the second flow occurs after time $x > 0$. For simplicity, it is assumed that the distribution functions (1.21) is absolutely continuous.

The renewal moments τ_n, $n \geq 1$, of the superposition (1.22) can now be described by the Markov chain $(\delta_n, \kappa_n; n \geq 0)$ taking values in the phase space E and then given by the transition probabilities

$$
\begin{aligned}
P_{11}(x, dy) \;:=&\; \mathcal{P}\{\delta_{n+1} = 1, \kappa_{n+1} \in dy | \delta_n = 1, \kappa_n = x\} \\
=&\; P_1(x - dy),
\end{aligned}
$$

$$
\begin{aligned}
P_{12}(x, dy) \;=&\; \mathcal{P}\{\delta_{n+1} = 2, \kappa_{n+1} \in dy | \delta_n = 1, \kappa_n = x\} \\
=&\; P_1(x + dy),
\end{aligned}
$$

$$
\begin{aligned}
P_{21}(x, dy) \;=&\; \mathcal{P}\{\delta_{n+1} = 1, \kappa_{n+1} \in dy | \delta_n = 2, \kappa_n = x\} \\
=&\; P_1(x + dy),
\end{aligned}
$$

$$
\begin{aligned}
P_{22}(x, dy) \;=&\; \mathcal{P}\{\delta_{n+1} = 2, \kappa_{n+1} \in dy | \delta_n = 2, \kappa_n = x\} \\
=&\; P_1(x - dy).
\end{aligned}
\tag{1.2.24}
$$

Note that the transition event

$$
1x \to 1dy
$$

means that $\theta^{(1)} \in x - dy$, that is,

$$
x - y < \theta^{(1)} \le x - y + dy.
$$

Similarly, the transition event

$$
1x \to 2dy
$$

means that $\theta^{(1)} \in x + dy$, that is,

$$
x + y < \theta^{(1)} \le x + y + dy.
$$

A remarkable peculiarity of the Markov chain $(\delta_n, \kappa_n; \; n \ge 0)$ with the transition probabilities (1.24) is ergodicity with the following stationary distribution

$$
\rho_1(dx) = \rho \bar{P}_2(x) dx,
$$
$$
\rho_2(dx) = \rho \bar{P}_1(x) dx,
\tag{1.25}
$$

where

$$
\rho = [m_1 + m_2]^{-1},
$$

$$m_i = E\theta^{(i)} = \int_0^\infty \bar{P}_i(x)dx, \quad i = 1, 2.$$

Further (Section 3.7) the procedure of superposition of renewal processes will be used in heuristic principles of phase merging algorithms.

1.2.5 The alternating renewal process

The alternating renewal process $(\kappa_n, \tau_n; n \geq 0)$ is described by the simplest general model of stochastic systems with two phase-states which distinguish between "working" and "repairing" states of system. The first component κ_n, $n \geq 0$, describes two states of the system: $\kappa_n = 1$ or 0 and the second component is the point process τ_n, $n \geq 0$, which determines the renewal times for the change of states. Consequently the distribution functions of renewal times are dependent on the state of the system:

$$G_k(t) := \mathcal{P}\{\theta_{n+1} \leq t/\kappa_n = k\}, \quad k = 1, 0.$$

In addition, the change of states is given by the stochastic matrix

$$P = \begin{bmatrix} 0 & 1 \\ 1 & 0 \end{bmatrix}.$$

Therefore the transitional probabilities are the following:

$$p_{10} = p_{01} = 1.$$

This two-component process $(\kappa_n, \tau_n; n \geq 0)$ is a Markov chain with the transitional probability matrix

$$Q(t) = [Q_{ij}(t); i, j = 1, 0]$$

with elements

$$Q_{10}(t) := \mathcal{P}\{\kappa_{n+1} = 0, \ \theta_{n+1} \leq t/\kappa_n = 1\} = G_1(t),$$

$$Q_{01}(t) := \mathcal{P}\{\kappa_{n+1} = 1, \ \theta_{n+1} \leq t/\kappa_n = 0\} = G_0(t), \quad (1.26)$$

and

$$Q_{00}(t) \equiv Q_{11}(t) \equiv 0.$$

The two component Markov process $(\kappa_n, \tau_n; n \geq 0)$ is called the *Markov renewal process*.

The general form of such a process is considered in Chapter 2. The evolution of the alternating renewal system in continuous time is described by the *semi-Markov process* $\kappa(t) := \kappa_{\nu(t)}$ where $\nu(t) := \max\{n : \tau_n \leq t\}$ is a *counting process* of the Markov renewal process $(\kappa_n, \tau_n; n \geq 0)$.

Let us now introduce the two-component Markov process in continuous time $(\kappa(t), \theta^+(t), \quad t \geq 0)$ where $\theta^+(t) := \tau_{\nu(t)+1} - t$ is the remaining time. In what follows we will use the stationary distribution of this process.

$$\lim_{t \to \infty} \mathcal{P}\{\kappa(t) = 1, \theta^+(t) \leq u\} = \rho_1 G_1^*(u),$$

$$\lim_{t \to \infty} \mathcal{P}\{\kappa(t) = 0, \theta^+(t) \leq u\} = \rho_0 G_0^*(u), \qquad (1.27)$$

where

$$\rho_k = m_k/m, \quad k = 1, 0,$$

$$m = m_1 + m_0,$$

$$m_k = \int_0^\infty \bar{G}_k(t)dt, \quad k = 1, 0, \qquad (1.28)$$

and

$$G_k^*(u) := \int_0^u \bar{G}_k(t)dt/m_k, \quad k = 1, 0, \qquad (1.29)$$

are the distribution of the stationary renewal times.

Consequently, the alternating renewal process in the steady-state regime with probability ρ_1 is situated in working state 1 during the stationary sojourn time θ_1^* and with probability ρ_0 is situated in repairing state 0 during stationary sojourn time θ_0^* with distributions (1.29).

Chapter 2

Markov renewal processes

Markov renewal processes (MRP) are quite a simple starting point for construction of a very wide class of jump Markov and semi-Markov processes.

Under some natural requirements on regularity the trajectories of Markov jump processes (MJP) behave as follows: an MJP occupies each state during a random time and has exponential distribution with some parameter depending on the state. Then a jump into a new state occurs according to the transition probabilities of a Markov chain with discrete time. Thus the MJP can be constructively defined by a stochastic kernel which determines the transition probabilities of the Markov chain and by non-negative function on states, which gives the intensities of exponential distributed sojourn times of the MJP in the states between two consecutive jumps.

The natural generalization of this construction of the MJP is considered as a semi-Markov process, for which the change of the states is also described by a Markov chain, but the sojourn times between two consecutive jumps are determined by some arbitrary distribution functions dependent on the state which has been visited as well as the next state to be entered.

Probabilistic structure of an MRP permits us to determine the semi-Markov process constructively by means of some characteristics which can be dealt with analytically.

2.1   Definition of Markov renewal process

Let $(E, \mathcal{E})$ be a measurable space. E is interpreted as a phase space of states of a stochastic system, $\mathcal{E}$ is a σ – algebra of subsets from E which can be regarded as a totality of all observable subsets of states of the system. $\mathcal{E}$ is supposed to contain one-point sets.

Definition 2.1. A *stochastic kernel* in a measurable space $(E, \mathcal{E})$ is given by a real-valued function P (x, B), $x \in E$, $B \in \mathcal{E}$ satisfying the following conditions: a) for a fixed x function $P(x, B)$ is a probability distribution on B : $P(x, E) \equiv 1$, b) P (x, B) is an $\mathcal{E}$-measurable function on x for a fixed B.

Remark 2.1. If $P(x, E) \leq 1$, then the kernel is s aid to be *substochastic*.

Stochastic kernel $P(x, B)$ determines transition (jump) probabilities from the state x into the set of states B for the stochastic system under consideration.

Example 2.1. In a discrete phase space $E = \{1, 2, \cdots, N\}$ stochastic matrix

$$P = [p_{ij}; i, j \in E]$$

all elements of which are non-negative: $p_{ij} \geq 0$ and $\sum_{j \in E} p_{ij} = 1$ for all $i \in E$.

Example 2.2. Let us have a probability distribution $P(B)$ on the real line $R = E$. Then the stochastic kernel $P(x, B) := P(B - x)$ defines a random walk on R:

$$\kappa_n = \sum_{k=0}^{n} \alpha_k, \; n \geq 0$$

where α_k, $k \geq 1$ are independent identically distributed random variables with the distribution function $P(B) = \mathcal{P}\{\alpha_k \in B\}$.

2.1.1 Homogeneous Markov chain

Let $(\Omega, \mathcal{F}, P)$ be a complete probability space, Ω is a sample space, $\mathcal{F}$ is a σ-algebra of subsets of Ω, and $\mathcal{P}$ is a probability measure on $\mathcal{F}$. More often, especially in applied problems, a sample space Ω is the Euclidean space R^N with finite dimension N. Then $\mathcal{F}$ is the σ-algebra of Borel sets of R^N.

Definition 2.2 A *homogeneous Markov chain* $(\kappa_n; n \geq 0)$ in a phase state space $(E, \mathcal{E})$ is defined as a sequence of random variables on the probability space $(\Omega, \mathcal{F}, \mathcal{P})$ with the transition probabilities $P(x, B) = \mathcal{P}\{\kappa_{n+1} \in B/\kappa_n = x\}$ and with an initial distribution $P_0(B) = \mathcal{P}\{\kappa_0 \in B\}$. The homogeneity of the Markov chain means that the on e-step transition probabilities do not depend on the passage time n.

The *Markov property* of Markov chain means that the joint distributions o f the chain's states are determined only by means of the initial distribution and of the one-step transition probabilities:

$$P\{\kappa_n \in B_n, \ \kappa_{n-1} \in B_{n-1}, \cdots, \kappa_0 \in B_0\}$$

$$= \ \int_{B_0} P(dx_0) \int_{B_1} P(x_0, \ dx_1) \cdots \int_{B_{n-1}} P(x_{n-2}, /dx_{n-1}) P(x_{n-1}, \ B_n).$$

According to the Markov property, the n-step transition probabilities

$$P_n(x, \ B) = \mathcal{P}\{\kappa_n \in B/\kappa_0 = x\}$$

are defined by the recurrent formula

$$P_n(x, \ B) = \int_E P_{n-1}(x, \ dy) P(y, B).$$

In addition, evidently $P_1(x, \ B) = P(x, B)$. The Markov family of stochastic kernels $P_n(x, \ B)$, $n \geq 1$, satisfies the *Chapman–Kolmogorov equation*

$$P_{n+m}(x, \ B) = \int_E P_n(x, \ dy) P_m(y, \ B)$$

which is an analytical expression for the Markov property of the Markov chain: given the state of the chain in the present

moment of time, the probability law of the future evolution of the system does not depend on the states occupied by the system in the past.

2.1.2   Semi-Markov kernel

Some special stochastic kernel is used for the definition of Markov renewal process.

Definition 2.3. A positive-valued function $Q(x, B, t)$, $x \in E$, $B \in \mathcal{E}$, $t \geq 0$, is said to be a *semi-Markov kernel* if the following condition s are satisfied:

a) $Q(x, B, t)$ is a substochastic kernel on a measurable space $(E, \mathcal{E})$ for a fixed $t > 0$: $Q(x, B, t) \leq 1$;

b) $Q(x, B, t)$ is a non-decreasing right-continuous function on $t \geq 0$ for a fixed x, B and $Q(x, B, 0) = 0$;

c) $Q(x, B, +\infty) =: P(x, B)$ is a stochastic kernel;

d) $Q(x, E, t) = G_x(t)$ is a distribution function on $t \geq 0$ for a fixed x.

In a discrete phase space $E = \{1, 2, \cdots, N\}$ a semi-Markov kernel is defined by a *semi-Markov matrix* $Q(t) = [Q_{ij}(t); \ i, j \in E]$ where $Q_{ij}(t)$ are non-decreasing functions of t and

$$P := Q(+\infty) = [p_{ij} := Q_{ij}(+\infty); \ i, \ j \in E]$$

is a stochastic matrix.

2.1.3   Markov Renewal Process

The ordinary renewal process, as it is well known, is given by a sequence of renewal moments

$$\tau_n = \sum_{k=1}^{n} \theta_n, \quad n \geq 1,$$

with independent identically distributed random variables θ_k, $k \geq 1$. It is natural to consider random variables θ_k, $k \geq 1$, which are dependent on some Markov chain $(\kappa_n; n \geq 0)$.

Definition 2.4. A homogeneous two-component Markov chain

$$(\kappa_n, \theta_n; \ n \geq 0)$$

taking values in $E \times [0, +\infty)$ is said to be a *Markov renewal process* (MRP) with phase space $(E, \mathcal{E})$ provided that its transition probabilities are given by a semi-Markov kernel:

$$Q(x, B, t) = \mathcal{P}\{\kappa_{n+1} \in B, \ \theta_{n+1} \leq t / \kappa_n = x\}.$$

The definition of MRP implies that the transition probabilities do not depend on the second component. This fact distinguishes a MRP from an arbitrary two-component Markov chain with a nonnegative second component. It follows at once from the definition of MRP that the first component of the MRP $(\kappa_n; n \geq 0)$ forms a Markov chain which is called an *imbedded Markov chain* (IMC).

One can obtain transition probabilities of the IMC by putting $t = +\infty$ in the semi-Markov kernel:

$$P(x, \; B) := Q(x, B, +\infty) = \mathcal{P}\{\kappa_{n+1} \in B/\kappa_n = x\}.$$

The nonnegative random variables θ_n, $n \geq 1$, called *renewal times*, define the intervals between the *Markov renewal moments*

$$\theta_n := \tau_n - \tau_{n-1}, n \geq 1,$$

$$\tau_n := \sum_{k=1}^{n} \theta_k.$$

The renewal moments τ_n, $n \geq 1$ taken together with the IM C $\kappa_n, n \geq 0$ form a two-component Markov chain $(\kappa_n, \; \tau_n; \; n \geq 0)$ which is said to be homogeneous in the second component [8].

The conditional distribution function of renewal time depends on the states of the IMC in the following way:

$$G_x(t) := Q(x, \; E, \; t) = \mathcal{P}\{\theta_{n+1} \leq t/\kappa_n = x\}$$

It is convenient to denote by θ_x the *sojourn time in state* $x \in E$ with the distribution function $G_x(t)$

$$\mathcal{P}\{\theta_x \leq t\} := G_x(t) = \mathcal{P}\{\theta_{n+1} \leq t/\kappa_n = x\}.$$

It will be always clear from the context what we mean: either the sojourn time θ_x in the state $x \in E$, or the renewal time θ_{n+1}
after the n-th step, assuming that $\kappa_n = x$.

Since $Q(x, \; B, \; t) \leq P(x, \; B)$ for any x, t and for all $B \in \mathcal{E}$ then the measure Q over $\mathcal{E}$ is absolutely
continuous   with   respect   to   measure P
and there exists such a measurable function $G_{xy}(t)$ that

$$Q(x, \; B, \; t) = \int_B G_{xy}(t) P(x, dy).$$

This implies the following probabilistic interpretation for the distribution function

$$G_{xy}(t) = \mathcal{P}\{\theta_{n+1} \le t/\kappa_n = x, \ \kappa_{n+1} = y\}.$$

The renewal times θ_n, $n \ge 1$ are *conditionally independent random variables on a fixed trajectory of the* IMC:

$$\mathcal{P}\{\theta_1 \le t_1, \cdots, \theta_n \le t_n/\kappa_0 = y_0, \cdots, \kappa_n = y_n\} = \prod_{k=1}^{n} G_{y_{k-1}y_k}(t_k).$$

Particularly, if $G_{xy}(t) = G_x(t)$ is independent from y, then the components of the MRP κ_n and θ_n are conditionally independent:

$$\mathcal{P}\{\kappa_{n+1} \in B, \ \theta_{n+1} \le t/\kappa_n = x\} =$$

$$= \mathcal{P}\{\kappa_{n+1} \in B/\kappa_n = x\} \times \mathcal{P}\{\theta_{n+1} \le t/\kappa_n = x\}$$

and the semi-Markov kernel has the following representation

$$Q(x, \ B, \ t) = P(x, \ B)G_x(t)$$

In a discrete phase space $E = \{1, 2, \cdots, N\}$ the MRP is determined by a semi-Markov matrix $Q(t) = [Q_{ij}(t); i, j, \in E]$ and, similarly in the case of a measurable phase space, $Q(t)$ can be represented in the form

$$Q_{ij}(t) = p_{ij}G_{ij}(t)$$

for all i, $j \in E$ such that $p_{ij} > 0$. In particular case of conditionally independent components of the MRP there is the following representation:

$$Q_{ij}(t) = p_{ij}G_i(t)$$

which is useful in analytical investigations.

2.1.4 Markov jump processes

A Markov jump process 'without aftereffect' $(\kappa_n, \theta_n; \ n \ge 0)$ is defined by the semi-Markov kernel with exponential distributions of renewal times:

$$Q(x, B, t) = P(x, B)(1 - e^{-q(x)t}).$$

Function $q(x) \geq 0$, $x \in E$, defines the intensities of renewal times:

$$\mathcal{P}\{\theta_{n+1} > t/\kappa_n = x\} = e^{-q(x)t}, \quad x \in E.$$

In a discrete phase space $E = \{1, 2, \ldots, N\}$ a semi-Markov kernel is given by a stochastic matrix $P = [p_{ij}; \; i, j \in E]$:

$$p_{ij} = \mathcal{P}\{\kappa_{n+1} = j/\kappa_n = i\}, \quad i, j \in E,$$

and a vector $q = (q_i; i \in E)$ with nonnegative components:

$$\mathcal{P}\{\theta_i > t\} = e^{-q_i t}, \quad i \in E.$$

Therefore,

$$Q_{ij}(t) = p_{ij}(1 - e^{-q_i t}), \quad i, j \in E.$$

A *regular homogeneous Markov jump process* is defined by the relation

$$x(t) = \kappa_{\nu(t)}.$$

The *counting process* is introduced by the relation

$$\nu(t) := \max\{n : \tau_n \leq t\}, \quad t > 0,$$

where $\tau_n := \sum_{k=1}^{n} \theta_k$, $n \geq 1$, $\tau_0 = 0$. The counting process $\nu(t)$ defines a number of renewal moments τ_n on the segment $(0, t]$:

$$\nu(t) = n, \; \tau_n \leq t < \tau_{n+1}$$

The counting process keeps constant values on the semi-intervals $[\tau_n, \tau_{n+1})$ and it is right continuous.

By definition, the components of the Markov renewal process without aftereffect are continuously independent.

To set a Markov jump process constructively, it is sufficient to have a stochastic kernel $P(x, B)$ which determines the transition probabilities of the imbedded Markov chain κ_n, $n \geq 0$, and a nonnegative function $q(x)$ which gives the intensities of sojourn times in states, and an initial distribution as well

$$\rho(B) = \mathcal{P}\{\kappa_0 \in B\}, \quad B \in \mathcal{X}.$$

The analytical approach to the theory of Markov jump processes is based on the *generating kernel* [4]

$$Q(x, B) = q(x)[P(x, B) - 1], \quad x \in X, \quad B \in \mathcal{X},$$

or, in a discrete phase space, on the *generating matrix*

$$Q = [q_{ij}; \quad i, j \in E],$$

$$q_{ij} := q_i p_{ij}, \quad i \neq j, \quad q_{ii} = -q_i.$$

Therefore, the generating matrix has the following form:

$$Q = q[P - I]$$

where $q := [q_i \delta_{ij}; \ i, j \in E]$ is a diagonal matrix of intensities; $I = [\delta_{ij}; \ i, j \in E]$ is the identity matrix. Here, as usual,

$$\delta_{ij} = \begin{cases} 1, & i = j, \\ 0, & i \neq j, \end{cases}$$

is the Kronecker symbol.

2.1.5 Stochastic representation of MRP

In applications, the range of states of a stochastic system often is under influence of some acting finite random factors. The simulation modelling of such a system can be realized by a MRP with the finite number of passage events.

Let a MRP be considered on a finite states space $E = \{1, 2, \cdots, N\}$ and for every $i \in E$ let $E_i \subset E$ be the set of states which can be reached from state i in one step: $p_{ij} > 0$ for $j \in E_i$.

Introduce the positive-valued random variables α_{ij}, $i \in E$, $j \in E_i$, which determine the random times of acting random factors. We assume that the random variables α_{ij} are mutually independent and have continuous distribution functions

$$F_{ij}(t) = \mathcal{P}\{\alpha_{ij} \leq t\} = 1 - \exp[-a_{ij}(t)].$$

Definition 2.5. The MRP $(\kappa_n, \theta_n; n \geq 0)$ *with a finite number of acting random factors* is given by the following stochastic relations:

$$\theta_i = \wedge_{j \in E_i} \alpha_{ij} \quad \{i \to k\} = \{\theta_i = \alpha_{ik}\} \tag{2.1}$$

Here we are using the signs $\wedge$ for the minimum and $\to$ for the passage event.

The first formula in (2.1) means that the sojourn time in state $i \in E$ is given by the minimum independent random variables $\alpha_{ij}, j \in E_i$.

The second formula in (2.1) means that the passage event from state i to state k is realized when the minimum of random factors is α_{ik}.

Now we can calculate the semi-Markov matrix which determines the MRP $(\kappa_n, \theta_n; n \geq 0)$ with a finite number of acting random factors. It's well known, that for independent random variables α_{ij} in (2.1) we get

$$
\begin{aligned}
\mathcal{P}\{\theta_i > t\} &= \mathcal{P}\{\wedge_{j \in E_i} \alpha_{ij} > t\} \\
&= \prod_{j \in E_i} \mathcal{P}\{\alpha_{ij} > t\} \\
&= \prod_{j \in E_i} \exp[-a_{ij}(t)] \\
&= \exp[-a_i(t)],
\end{aligned}
$$

where by definition $a_i(t) := \sum_{j \in E_i} a_{ij}(t)$. Using the conditional expectation we can calculate

$$
\begin{aligned}
Q_{ik}(t) &:= \mathcal{P}\{\kappa_{n+1} = k, \theta_{n+1} \leq t/\kappa_n = i\} \\
&= \mathcal{P}\{\theta_i = \alpha_{ik} \leq t\} = \mathcal{P}\{\alpha_{ik} < \wedge_{j \neq k} \alpha_{ij}; \theta_i \leq t\} \\
&= \int_0^t e^{-a_{ik}(s)} da_{ik}(s) \exp \sum_{j \neq k} a_{ij}(s).
\end{aligned}
$$

Therefore we have such a representation of the semi-Markov matrix elements:

$$Q_{ik}(t) = \int_0^t e^{-a_i(s)} da_{ik}(s), \quad i, k \in E.$$

The transition probabilities of the IMC (κ_n; $n \geq 0$) have the following form:

$$p_{ij} = Q_{ij}(+\infty) = \int\limits_0^\infty e^{-a_i(s)} da_{ij}(s).$$

Remark 2.2. The stochastic representation (2.1) doesn't imply that $\theta_{ij} = \alpha_{ij}$, since $G_{ij}(t) = Q_{ij}(t)p_{ij} \neq F_{ij}(t)$ even in the case of exponentially distributed factors α_{ij}.

2.2   Semi-Markov processes

Let ($\kappa_n, \theta_n; n \geq 0$) be a Markov renewal process with a measurable phase space $(E, \mathcal{E})$ and a semi-Markov kernel $Q(x, B, t)$.

Definition 2.6. A *semi-Markov process* (SMP) $\kappa(t)$, $t \geq 0$ is determined by the relation

$$\kappa(t) := \kappa_{\nu(t)}$$

the SMP $\kappa(t)$ also keeps constant values on the semi-intervals $[\tau_n, \tau_{n+1})$ and is right-continuous:

$$\kappa(t) = \kappa_n, \ \tau_n \leq t < \tau_{n+1}$$

In addition

$$\kappa(\tau_n) = \kappa_n, \ n \geq 0$$

This relation illustrates the notion of the imbedded Markov chain (κ_n; $n \geq 0$).

For the SMP $\kappa(t)$ the renewal times $\theta_{n+1} = \tau_{n+1} - \tau_n$ can be naturally interpreted as *sojourn times* in the states κ_n.

As it follows we shall consider only *regular* SMP which have a finite number of renewals during the finite time interval with probability one:

$$\mathcal{P}\{\nu(t) < +\infty\} = 1 \text{ for all } t > 0.$$

2.2.1 Markov renewal equation

We introduce the convolution of a semi-Markov kernel $Q(x,\ B,\ t)$ with some given real valued function $U(t,\ x)$ in such a form:

$$Q * U := \int_0^t \int_E Q(x,\ dy,\ ds)U(t - s,\ y)$$

The convolution of a semi-Markov kernel is really an integral convolution in time variable t and integral operation over phase variable x.

In a discrete phase space $E = \{1,\ 2, \cdots, N\}$ convolution of matrix $Q(t) = [Q_{ij}(t);\ i,\ j \in E]$ with some vector-function $U(t) = (U_i(t);\ i \in E)$ is matrix operation of phase component:

$$Q * U := (\sum_{j \in E} \int_0^t Q_{ij}(ds)U_j(t - s);\ i \in E)$$

Note that the n-th convolution of a semi-Markov kernel itself

$$Q^{(n)}(x,\ B,\ t) := \int_0^t \int_E Q(x,\ dy,\ ds)Q^{(n-1)}(y,\ B,\ t - s),\ n \geq 2$$

gives us the distribution of the MRP:

$$Q^{(n)}(x,\ B,\ t) = \mathcal{P}\{\kappa_n \in B,\ \tau_n \leq t / \kappa_0 = x\}$$

The main object in analysis is the integral *Markov renewal equation* (MRE) which can be determined by a semi-Markov kernel $Q(x,\ B,\ t)$ and some given real valued function $V(t,\ x)$.

Definition 2.7. A *Markov renewal equation* with given function $V(t,\ x)$ has the following form

$$U(t,\ x) - \int_0^t \int_E Q(x,\ dy,\ ds)U(t - s,\ y) = V(t,\ x)$$

Using the sign of convolution $*$ the MRE can be represented as follows:

$$[I - Q] * U = V,$$

where I is identity operator: $I * U = U * I = U$.

Considering the semi-Markov kernel $Q(t)$ as an operator in the normed linear (Banach) space $\mathcal{B}_E$ of bounded real-valued functions $U(x)$, $x \in E$ with sup-norm: $\|U(x)\| := \sup_{x \in E} |U(x)|$ we can write down the MRE in such a form:

$$U(t) - \int_0^t Q(ds)U(t-s) = V(t), \ t \geq 0,$$

where function $U(t)$ and $V(t)$ take values in the normed space $\mathcal{B}_E$. In particular, when E contains one state then the MRE is the renewal equation which is well-known in the renewal theory [5]. The integral convolution over t gives us an opportunity to apply the Laplace transformation to the MRE ($Re\lambda > 0$) :

$$[I - \tilde{Q}(\lambda)]\tilde{U}(\lambda) = \tilde{V}(\lambda),$$

where

$$\tilde{U}(\lambda) := \int_0^\infty e^{-\lambda t}U(t)dt, \ \tilde{Q}(\lambda) := \int_0^\infty e^{-\lambda t}Q(dt).$$

Note, that the integral operator $\tilde{Q}(\lambda)$ has non-negative contractive kernel: $0 \leq \tilde{Q}(x, B, \lambda) \leq 1$ when $Re\lambda > 0$ and $\tilde{Q}(x, B, 0) = P(x, B)$ is a stochastic kernel.

2.2.2   Characteristics of MRP

The various functionals of MRP and SMP can be determined as a solution of the MRE. We consider only some of them which will be used later on.

Lemma 2.1. The transition probabilities of the semi-Markov process

$$\varphi(t, x; B) := \mathcal{P}\{\kappa(t) \in B/\kappa(0) = x\}$$

is the solution of the following MRE

$$\varphi(t, x; B) - \int_0^t \int_E Q(x, dy, ds)\varphi(t-s, y; B) = I_B(x)\bar{G}_x(t),$$

where

$$I_B(x) := \begin{cases} 1, & x \in B, \\ 0, & x \notin B, \end{cases}$$

and $\bar{G}_x(t) := 1 - G_x(t)$.

Proof of Lemma 2.1. Putting $\tau_1 = \theta_x$ under condition $\kappa_0 = x$ we get

$$\mathcal{P}\{\kappa(t) \in B / \kappa_0 = x\} = \mathcal{P}\{\kappa(t) \in B, \ \theta_x > t / \kappa_0 = x\}$$
$$+ \mathcal{P}\{\kappa(t) \in B, \ \theta_x \leq t / \kappa_0 = x\}$$

It is easy to calculate the first term

$$\mathcal{P}\{\kappa(t) \in B, \ \theta_x > t / |\kappa_0 = x\} = I_B(x)\bar{G}_x(t)$$

Using the Markov property of the semi-Markov process at the renewal mo-
lu₋ ⁻ⁿ set the second term in such a form:

$$\mathcal{P}\{\kappa(t) \in B, \ \theta_x \leq t / \kappa_0 = x\} = \int_0^t \int_E Q(x, \ dy, \ ds)\varphi(t - s, \ y; \ B)$$

So, adding two terms we get the MRE for the transition probabilities of
the SMP.

Lemma 2.2. The average value of the SMP

$$\varphi(t, \ x) := E[f(\kappa(t))/\kappa(0) = x]$$

for some bounded real-valued function $f(x)$, $x \in E$ is the solution of the
MPE in such a form

$$\varphi(t, \ x) - \int_0^t \int_E Q(x, \ dy, \ ds)\varphi(t - s, \ y) = f(x)\bar{G}_x(t)$$

Proof of Lemma 2.2. In order to construct the MRE for the ave rage
value of the SMP we have to calculate only the right-hand side MRE, i. e.

$$V(t, \ x) := E[f(\kappa(t)), \ \theta_x > t / \kappa_0 = x] = f(x)\bar{G}_x(t).$$

The useful characteristic of the SMP, which has various applications, is the sojourn time in some subset of states, $E_0 \subset E$. Introduce the random variable

$$\zeta_x^{E_0} := \min\{t : \kappa(t) \notin E_0 / \kappa_0 = x\}, \quad x \in E_0.$$

Lemma 2.3. The distribution function of the sojourn time in subset of states

$$\varphi_0(t, \, x) := \mathcal{P}\{\zeta_x^{E_0} > t\}$$

satisfies the following MRE

$$\varphi_0(t, \, x) - \int\limits_0^t \int_{E_0} Q(x, \, dy, \, ds)\varphi_0(t - s, \, y) = \bar{G}_x(t), \ x \in E_0.$$

Proof of Lemma 2.3. To be sure that this equation is valid we have to calculate only the right-hand side at MRE, which has the following form

$$\mathcal{P}\{\zeta_x^{E_0} > t, \, \theta_x > t / \kappa_0 = x\} = \mathcal{P}\{\theta_x > t\} = \bar{G}_x(t).$$

2.2.3   Auxiliary processes

The Markov renewal process $(\kappa_n, \tau_n; n \geq 0)$ induces some auxiliary processes which has various applications. At first we introduce the *point process*

$$\tau(t) := \tau_{\nu(t)}, \quad t \geq 0.$$

which is fixing renewal moments $\tau_n, \ n \geq 0$.

Definition 2.8. *Auxiliary processes associated with a Mark ov renewal process $(\kappa_n, \ \tau_n; \ n \geq 0)$ are defined as follows:*

$\theta(t) := \theta_{\nu(t)+1}$ is a *sojourn time process in stat e $\kappa(t)$*;
$\theta^-(t) := t - \tau(t)$ is the *previous sojourn time (de fect-process)*;
$\theta^+(t) := \theta(t) - \theta^-(t)$ is the *remaining sojourn time (excess-process)*.

The auxiliary processes have the *regenerative property* with re spect to the MRP. Introduce the filtration of σ-algebras associated with the MRP and the SMP:

$$F_n := \sigma\{\kappa_0, \ \kappa_1, \ \tau_1, \cdots, \kappa_n, \ \tau_n\}, \ n \geq 0;$$
$$F_t := \sigma\{\kappa(s); \ 0 \leq s \leq t\}, \ t \geq 0;$$

$$F_t^+ := \sigma\{\kappa(s), \ \theta(s); \ 0 \leq s \leq t\}, \ t \geq 0.$$

Note that the renewal moments τ_n are Markovian moments with respect to F_t [15]. Moreover $F_{\tau_n} = F_n$. The auxiliary processes $\theta(t)$ and $\theta^+(t)$ are F_t^+-measurable whereas $\theta^-(t)$ and $\tau(t)$ are F_t-measurable. Every auxiliary process $\theta(t)$ and $\theta^\pm(t)$ extends the SMP $\kappa(t)$ to a Markov process: for the Borel bounded function $f(t, \ x)$, $x \in E$, $t \geq 0$ and all $s \leq t$ the *Markov property holds*: [14]

$$E[f(\theta(t), \ \kappa(t))/F_s^+] = E[f(\theta(t), \ \kappa(t))/\theta(s), \ \kappa(s)].$$

This Markov property remains true if $\theta(t)$ is substituted by $\theta^\pm(t)$.

2.2.4 Regenerative processes

The auxiliary processes posses a useful regenerative property. Let $\zeta(t)$, $t \geq 0$ be a real-valued stochastic process with right-continuous trajectories having left limits and

$$F_t^\zeta := \sigma\{\zeta(s); \ 0 \leq s \leq t\}$$

be the natural filtration of σ-algebras generated by trajectories of $\zeta(t)$.

Definition 2.9. A stochastic process $(\zeta(t), \ F_t^\zeta; \ t \geq 0)$ is called *regenerative with respect to the* MRP $(\kappa_n, \ \tau_n; \ n \geq 0)$ if the following conditions are valid:
a) the renewal moments τ_n are Markovian moments with respect to F_t^ζ;
b) the imbedded Markov chain κ_n is $F_{\tau_n}^\zeta$ – measurable;
c) The Markovian property is fulfilled in such a form:

$$E[f(\kappa(\tau_n + t), \ \zeta(\tau_n + t))/F_{\tau_n}^\zeta] = E[f(\kappa(t), \ \zeta(t))/\kappa_n] \qquad (2.2)$$

It is easy to verify that auxiliary processes possess such a regenerative property with respect to the MRP $(\kappa_n, \ \tau_n; \ n \geq 0)$ which induced them.

The conditional distribution of a regenerative process to the first renewal moment

$$\varphi(t, \ x, \ B) := \mathcal{P}\{\zeta(t) \in B, \tau_1 > t/\kappa_0 = x\}$$

determines the distribution of the process $\zeta(t)$ on an arbitrary finite

time interval. Indeed, the regenerative property (2.2) together with the definition of the conditional distribution give us such a relation

$$\mathcal{P}\{\zeta(\tau_n + t) \in B, \ \tau_{n+1} > t + \tau_n/\kappa_n = x\} = \varphi(t, \ x, \ B).$$

The conditional distribution of the regenerative process satisfies the Markov renewal equation. Consider, for example, the defect-process $\theta^-(t) := t - \tau(t)$.

Lemma 2.4. The distribution function of the defect-process

$$\varphi^-(t, \ x; \ u) := \mathcal{P}\{\theta^-(t) > u/\kappa_0 = x\}$$

is the solution of the following MRE

$$\varphi^-(t, \ x; \ u) - \int_0^t \int_E \varphi(x, \ dy, \ ds)\varphi^-(t - s, \ y, \ u) = I(t > u)\bar{G}_x(t)$$

where

$$I(t > u) := \left\{ \begin{array}{ll} 1, & t > u \\ 0, & t \le u \end{array} \right.$$

is the indicator-function.

Proof of Lemma 2.4. We have to calculate the right-hand part of the MRE. So, we set

$$V^-(t, \ x) := \mathcal{P}\{\theta^-(t) > u, \ \theta_x > u/\kappa_0 = x\} = I(t > u)\bar{G}_x(t).$$

The main problem for a regenerative processes is to investigate the ergodic property of such a processes [18].

2.3 Ergodicity and stationary distribution

The phase merging and averaging algorithms are based on the ergodic property of Markov processes.

2.3.1 Ergodic property of Markov chain

Let $(\kappa_n;\ n \geq 0)$ be a Markov chain induced by a stochastic kernel $P(x,\ B)$ on a standard phase space $(E,\ \mathcal{E})$.

Ergodic property of Markov chain means that there exists

$$\lim_{n\to\infty} \frac{1}{n} \sum_{k=0}^{n-1} \varphi(\kappa_k) = \int_E \rho(dx)\varphi(x)$$

for all bounded functions $\varphi \in \mathcal{B}_E$. Here $\rho(dx)$ is called the *stationary distribution* of Markov chain:

$$\rho(B) = \int_E \rho(dx)P(x,\ B),\ B \in \mathcal{E},\ \rho(E) = 1. \qquad (2.3)$$

Choosing the initial distribution $\mathcal{P}\{\kappa_0 \in B\} = \rho(B)$ we immediately get from (2.3) that $\mathcal{P}\{\kappa_n \in B\} = \rho(B)$ for all $n \geq 1$. This explains the notion of stationary distribution. There are various well-known criteria of ergodicity of a Markov chain [15]. We will consider here only *aperiodic uniformly ergodic Markov chains:*

$$\lim_{n\to\infty} \mathcal{P}\{\kappa_n \in B/\kappa_0 = x\} = \rho(B)$$

uniformly on $x \in E$ and $B \in \mathcal{E}$. Also we assume that the ergodic Markov chain is *irreducible* [15]: for every $B \in \mathcal{E}$ with $\rho(B) > 0$ and for all $x \in E$ there exists $n \geq 1$ for which $\mathcal{P}\{\kappa_n \in B/\kappa_0 = x\} > 0$.

But in a phase merging algorithm we will also use the reducible ergodic Markov chain which is induced by a stochastic kernel $P(x,\ B)$, which is coordinated with splitting phase space

$$E = \cup_{k=1}^N E_k,\quad E_k \cap E_{k,} = \emptyset,\quad k \neq k'$$

in such a form

$$P(x,\ E_k) = 1_k(x) := \begin{cases} 1, & x \in E_k \\ 0, & x \notin E_k. \end{cases}$$

The classes of states E_k are closed sets for ergodic Markov chain. Ergodicity of reducible Markov chains means that there exist stationary distributions $\rho_k(dx),\ 1 \leq k \leq N$:

$$\rho_k(B) = \int_{E_k} \rho_k(dx)P(x,\ B),\ B \in \mathcal{E}_k,\ \rho_k(E_k) = 1$$

Here $\mathcal{E}_k$ is contraction of σ-algebra $\mathcal{E}$ on set E_k. Essentially, a reducible ergodic Markov chain consists of N separate irreducible ergodic Markov chains on classes $E_k,\ 1 \leq k \leq N$.

2.3.2   Ergodicity of semi-Markov process

Ergodicity of Markov chain $(\kappa_n;\ n \geq 0)$ gives us an opportunity to investigate ergodic property of a semi-Markov process $\kappa(t) = \kappa_{\nu(t)}$ and other auxiliary processes introduced in Section 2.2. It turns out that for this purpose the renewal limit theorem which we formulate here in certain simplified form, is very useful.

Renewal limit theorem [18]. Let the following conditions be satisfied:
a) the stochastic kernel $P(x,\ B) = Q(x,\ B,\ +\infty)$ induces irreducible ergodic Markov chain with the stationary distribution $\rho(B)$, $B \in \mathcal{E}$.
b) there exist bounded first moments of sojourn times

$$m(x) := \int_0^\infty \bar{G}_x(t)dt \leq C < +\infty$$

and the stationary average of sojourn time

$$m := \int_E \rho(dx)m(x) > 0.$$

c) the distribution functions $G_x(t) := Q(x,\ E,\ t)$, $x \in E$ of sojourn times are nonarithmetical [18].
d) a non-negative function $V(t,\ x)$ is *directly Riemann integrable* on $t \geq 0$ [5], so, there exists finite integral

$$\int_E \rho(dx) \int_0^\infty V(t,\ x)dt < \infty$$

Then there exists the unique solution $U(t,\ x)$ of the Markov renewal equation

$$U(t,\ x) - \int_0^t \int_E Q(x,\ dy,\ ds)U(t-s,\ y) = V(t,x)$$

and the following limiting result holds true:

$$\lim_{t \to \infty} U(t,\ x) = \int_E \rho(dx) \int_0^\infty V(t,\ x)dt/m.$$

Note, that a non-decreasing integrable function $V(t)$ is directly Riemann integrable [5].

The renewal limit theorem is a source of various limiting results for regenerati ve processes with respect to a Markov renewal process $(\kappa_n, \theta_n; n \geq 0)$ induced by a semi-Markov kernel $Q(x, B, t)$.

The transition probabilities of the semi-Markov process

$$\Phi(t, x; B) = \mathcal{P}\{\kappa(t) \in B/\kappa(0) = x\}$$

satisfy the Markov renewal equation (see Lemma 2.1) with the right-hand side $V(t, x) = I_B(x)\bar{G}_x(t)$. Applying the renewal limit theorem to this function we get such a limiting result

$$\lim_{t \to \infty} \mathcal{P}\{\kappa(t) \in B/\kappa(0) = x\} = \int_B \rho(dx)m(x)/m, \ B \in \mathcal{E}.$$

So, the stationary distribution of the semi-Markov process has the following representation

$$\pi(B) = \int_B \rho(dx)m(x)/m, \ B \in \mathcal{E},$$

where $\rho(dx)$ is the stationary distribution of the imbedded Markov chain.

In a discrete phase space $E = \{1, 2, \cdots, N\}$ the stationary distribution of the semi-Markov process is given by the relation

$$\pi_k = \rho_k m_k/m,$$

$$m = \sum_{k \in E} \rho_k m_k.$$

The distribution function of the auxiliary defect-process $\theta^-(t) = t - \tau(t)$ introduced in Section 2.2

$$\varphi^-(t, x; u) = \mathcal{P}\{\theta^-(t) > u/\kappa_0 = x\}$$

can be considered as the solution of the Markov renewal equation with right-hand side $V(t, x) = I(t > u)\bar{G}_x(t)$ (see Lemma 2.2).

Applying the renewal limit theorem to this function we get such a limiting result:

$$\lim_{t \to \infty} \mathcal{P}\{\theta^-(t) > u/\kappa_0 = x\} = \int_E \rho(dx) \int_u^\infty \bar{G}_x(t)dt/m.$$

Similarly one can obtain a more general limiting result for the common distribution function (see Lemma 2.3)

$$\lim_{t \to \infty} \mathcal{P}\{\kappa(t) \in B, \ \theta^-(t) > u/\kappa_0 = x\} = \int_B \rho(dx) \int_u^\infty \bar{G}_x(t)dt/m.$$

Using the relation between stationary distributions of imbedded Markov chain $\rho(dx)$ and semi-Markov process $\pi(dx)$

$$\pi(dx) = \rho(dx)m(x)/m,$$

the last limiting result can be rewritten as follows:

$$\lim_{r \to \infty} \mathcal{P}\{\kappa(t) \in B, \ \theta^-(t) > u/\kappa_0 = x\} = \int_B \pi(dx)\bar{G}_x^*(u)$$

where $\bar{G}_x^*(u) := \int_u^\infty \bar{G}_x(t)dt/m(x)$

is called the *stationary distribution of sojourn time in state* $x \in E$:
$\bar{G}_x^*(u) := \mathcal{P}\{\theta_x^* > u\}$.

Note, that this limiting result is more clearly expressed in a discrete phase space:

$$\lim_{t \to \infty} \mathcal{P}\{\kappa(t) = k, \ \theta^-(t) > u/\kappa_0 = r\} = \pi_k \bar{G}_k^*(u), \quad k \in E.$$

Statistical interpretation of stationary distribution. In heuristic principles of phase mergence given in Section 3.5 statistical interpretation of stationary distributions was used.

Uniform ergodicity of the Markov chain $(\kappa_n; \ n \geq 0)$ means that after some large enough moment n the Markov chain $(\kappa_{n+k}, \ k \geq 0)$ can be considered in *steady-state (stationary) regime*, so that

$$\mathcal{P}\{\kappa_{n+k} \in B\} \simeq \rho(B), \ k \geq 0.$$

Ergodic property of Markov chain means that for some large enough n

$$\frac{1}{n}\sum_{k=0}^{n-1} I_B(\kappa_n) \simeq \rho(B),$$

Therefore the stationary distribution $\rho(B)$ can be interpreted as part of sojourn time in the set of states B. Another useful interpretation of stationary distribution $\rho(B)$ is as follow s. Let $(\kappa_n^i,\ n \geq 0)$, $1 \leq i \leq N$ be a family of independent ergodic Markov chain with the same stationary distribution $\rho(B)$.

Then the number $\nu(B)$ of Markov chains situated in states of set B can be approximated as follows

$$\nu(B) \simeq N\rho(B)$$

or, in another form,

$$\rho(B) \simeq \nu(B)/N.$$

Chapter 3

Phase merging algorithms

The general scheme of phase merging which is described in Section 1.2, is now realized as *phase merging algorithms* (PMA) which can be applied to stochastic systems described by Markov and semi-Markov processes. We consider only two different PMA which are used in phase averaging algorithms and in the diffusion approximation of stochastic evolutions. In Chapter 4 we consider the phase merging with stoppings (absorption) and in reducible phase space.

3.1   Reducible-invertible operators

The main properties of the generating operator of homogeneous Markov processes are considered in abstract form [12].

Properties of reducible-invertible operators. We introduce some necessary notations and notions of the theory of operators in a complete Banach space.

As a rule, the normed linear space of vector-functions with argument, which takes values in the phase space E is denoted by $\mathcal{B}_E$ or, without subscript, $\mathcal{B}$. The dual space of linear functionals on $\mathcal{B}$ is denoted by $\mathcal{B}^*$. We usually denote a linear functional by $l(\varphi) \in \mathcal{B}^*$, where $\varphi \in \mathcal{B}$.

Let Q be a linear operator acting in $\mathcal{B}$. The conjugate to Q operator is denoted by Q^*:

$$Q^*l(\varphi) := l(Q\varphi).$$

43

In order to understand better the abstract notion of the theory of operators it is worth to consider a finite-dimensional Euclidean vector-space R^N that is composed of finite column-vectors $\varphi = (\varphi_k; 1 \leq k \leq N)$ with the norm defined by

$$\|\varphi\| := \sup_{1 \leq k \leq N} |\varphi_k|.$$

Then $\mathcal{B}^*$ is composed of row-vectors $l^* = (l_k; 1 \leq k \leq N)$ with the norm $\|l\| := \sum_{k=1}^{N} |l_k|$. The linear functional $l(\varphi) := \sum_{k=1}^{N} l_k \varphi_k$ is the scalar product of row-vector l^* and column-vector φ which we consider as product of two rectangular matrices l^* of size $1 \times N$ and φ of size $N \times 1$:

$$l^*\varphi := \sum_{k=1}^{N} l_k \varphi_k.$$

Notice, that the product of two matrices φ and l^* gives us the square matrix:

$$\varphi l^* = [\varphi_k l_r; \ 1 \leq k, r \leq N].$$

The bounded linear operator Q is determined by the matrix

$$Q = [q_{kr}; \ 1 \leq k, r \leq N]$$

which acts on a colomn-vector in usual manner:

$$Q\varphi = \psi,$$

$$\psi_k = \sum_{r=1}^{N} q_{kr} \varphi_r, \quad 1 \leq k \leq N.$$

Definition 3.1. A bounded linear operator Q is called *reducible-invertible* if the Banach space $\mathcal{B}$ can be decomposed in direct sum of two subspaces:

$$\mathcal{B} = N_Q \oplus R_Q, \tag{3.1}$$

where $N_Q := \{\varphi : Q\varphi = 0\}$, that is the *null-space* of the operator Q with a nontrivial (nonzero) dimension: $\dim N_Q \geq 1$. The subspace

$$R_Q := \{\psi : Q\varphi = \psi, \ \varphi \in \mathcal{B}\}$$

is called a *range* of values of an operator Q.

Decomposition (3.1) generates the projector Π on the subspace N_Q:

$$\Pi\varphi := \begin{cases} \varphi, & \varphi \in N_Q, \\ 0, & \varphi \in R_Q. \end{cases} \tag{3.2}$$

The operator $I - \Pi$ is the projector on the subspace R_Q

$$[I - \Pi]\varphi := \begin{cases} 0, & \varphi \in N_Q, \\ \varphi, & \varphi \in R_Q. \end{cases} \tag{3.3}$$

Example 3.1. The matrix $\Pi = [\varrho_r;\ 1 \leq k, r \leq N]$ with identical rows is the projector on the one-dimensional space $N = \{c\mathbf{1};\ c \in R\}$, where $\mathbf{1} := (1, 1 \leq k \leq N)$, that is the vector with identical elements equal to 1.

Definition 3.2. A bounded linear operator Q is called *normally resolvable* if the equation

$$Q\varphi = \psi \tag{3.4}$$

has a solution for all $\psi \in R_Q$.

Notice, that a reducible-invertible operator is always normally resolvable.

The important property of reducible-invertible operator is the existence of a *generalized inverse operator*.

Lemma 3.1. For a normally resolvable operator Q there exists the inverse operator $[Q + \Pi]^{-1}$.

Proof of Lemma 3.1. We apply the projector Π to the equation

$$[Q + \Pi]\ \varphi = \psi \tag{3.5}$$

we get the relation

$$\Pi\varphi = \Pi\psi.$$

Using the obvious relation $\Pi Q\varphi = Q\Pi\varphi = 0$, equation (3.5) can be rewritten in the form

$$Q\varphi = [I - \Pi]\ \psi.$$

But $[I - \Pi]\,\psi \in R_Q$ according to the definition of the projector $I - \Pi$. Hence, equation (3.4) with the normally resolvable operator Q has solution for all $\psi \in \mathcal{B}$.

Definition 3.3. The *potential operator* (or, simply, the *potential*) of the reducible-invertible operator Q is the operator

$$R_0 := [Q + \Pi]^{-1} - \Pi. \tag{3.6}$$

Notice, that it is easy to check that the potential can be represented in the following form:

$$R_0 := [Q - \Pi]^{-1} + \Pi.$$

The best understanding of the notion of potential comes with the example of matrix operator. According to definition (3.2) in some vector-basis the projector acts as an identity operator on N_Q and as a null-operator on R_Q so that it can be represented as a block-matrix:

$$\Pi = \begin{array}{c} N_Q \ R_Q \\ \begin{bmatrix} I & 0 \\ 0 & 0 \end{bmatrix} \begin{array}{c} N_Q \\ R_Q \end{array} \end{array}.$$

Analogously, it is possible to represent the operator Q with a nontrivial null-space N_Q in the block-matrix form:

$$Q = \begin{array}{c} N_Q \ R_Q \\ \begin{bmatrix} 0 & 0 \\ 0 & Q_0 \end{bmatrix} \begin{array}{c} N_Q \\ R_Q \end{array} \end{array}.$$

The matrix Q_0 acts in the subspace R_Q and according to definition 3.2 there exists an inverse matrix Q_0^{-1}. So we get

$$Q + \Pi = \begin{array}{c} N_Q \ R_Q \\ \begin{bmatrix} I & 0 \\ 0 & Q_0 \end{bmatrix} \begin{array}{c} N_Q \\ R_Q \end{array} \end{array}, \qquad [Q + \Pi]^{-1} = \begin{array}{c} N_Q \ R_Q \\ \begin{bmatrix} I & 0 \\ 0 & Q_0^{-1} \end{bmatrix} \begin{array}{c} N_Q \\ R_Q \end{array} \end{array}.$$

Finally, the potential R_0 has the representation

$$R_0 = [Q + \Pi]^{-1} - \Pi = \begin{array}{c} N_Q \ R_Q \\ \begin{bmatrix} 0 & 0 \\ 0 & Q_0^{-1} \end{bmatrix} \begin{array}{c} N_Q \\ R_Q \end{array} \end{array}.$$

Hence, R_0 is a generalized inverse operator to Q on the subspace R_Q. So the potential can be called *reducible-inverse* to Q.

The main properties of the potential R_0 are as follows:

(a) $QR_0 = R_0Q = I - \Pi$;

(b) $\Pi R_0 = R_0\Pi = 0$;

(c) $QR_0^n = R_0^n\Pi = R_0^{n-1}$;

(d) $\|R_0\| = \|Q_0^{-1}\|$.

The matrix representation of the potential is the best explanation of these properties. But it is easy to give formal proofs of them. For example, with the help of Lemma 3.1 we get

$$I - \Pi = Q[Q + \Pi]^{-1} = Q[Q + \Pi]^{-1} - Q\Pi = QR_0.$$

Proposition 3.1. The general solution of the equation with reducible-invertible operator Q

$$Q\varphi = \psi \tag{3.7}$$

under the solvability condition $\Pi\psi = 0$ admits the representation

$$\varphi = R_0\psi + \varphi_0, \quad \varphi_0 \in N_Q. \tag{3.8}$$

Notice, that under additional condition $\Pi\varphi = 0$ the equation (3.7) has a unique solution which can be represented in the form

$$\varphi = R_0\psi, \quad \Pi\varphi = 0. \tag{3.9}$$

This fact will be used in the construction of phase merging and averaging algorithms for stochastic systems.

3.2 Perturbation of reducible-invertible operators

Diverse problems of the asymptotical analysis of stochastic systems can be reduced to the problem of *singular perturbation of reducible-invertible operators* [12].

Let Q be a bounded reducible-invertible operator on the Banach space $\mathcal{B}$:

$$\mathcal{B} = N_Q \oplus R_Q,$$

$N_Q := \{\varphi : \ Q\varphi = 0\}$ is the null-space of Q with $\dim N_Q \geq 1$; $R_Q := \{\psi : Q\varphi = \psi, \varphi \in \mathcal{B}\}$ is the range of values of Q.

The *problem of asymptotic singular perturbation of a reducible-invertible operator* Q with small parameter $\varepsilon > 0$ and perturbing operator Q_1 is formulated in the following way. We have to construct the vector $\varphi^\varepsilon = \varphi + \varepsilon\varphi_1$ which realizes the asymptotic representation

$$[\varepsilon^{-1}Q + Q_1]\varphi^\varepsilon = \psi + \varepsilon\theta_\varepsilon \tag{3.10}$$

for some given vector ψ and with uniformly bounded in norm vector θ_ε: $\|\theta_\varepsilon\| \leq C$ as $\varepsilon \to 0$.

Such a problem is arising as the problem of asymptotic solution of the equation

$$[Q + \varepsilon Q_1]\varphi^\varepsilon = \psi^\varepsilon$$

for a given vector ψ^ε. The same problem is appearing when the operator inverse to a singular perturbed one is constructed

$$[Q + \varepsilon Q_1]^{-1} = \varepsilon^{-1}Q^0 + Q^1 + \cdots$$

The solutions of singular perturbation problems will be utilized in construction and verification of phase merging and averaging algorithms for stochastic systems. There exist many situations which can not be classified. But it is possible to extract some logically complete variants of these problems [12].

The solution of singular perturbation problems (3.10) is based on the properties of reducible-invertible operators which were given in Section 3.1.

The left-hand side of (3.10) can be represented in the following form.

$$[\varepsilon^{-1}Q + Q_1](\varphi + \varepsilon\varphi_1) = \varepsilon^{-1}Q\varphi + [Q\varphi_1 + Q_1\varphi] + \varepsilon Q_1\varphi_1.$$

In order to realize the right-hand side of the representation (3.10) we set

$$Q\varphi = 0, \quad Q\varphi_1 + Q_1\varphi = \psi, \quad Q_1\varphi_1 = \theta_\varepsilon. \tag{3.11}$$

The first equality in (3.11) means that $\varphi \in N_Q$. The third equality in (3.11) means that the vector $\theta_\varepsilon = Q_1\varphi_1$ is independent of ε. The main problem is to solve the equation

$$Q\varphi_1 = \psi - Q_1\varphi \tag{3.12}$$

which we get from the second equality of (3.11) with given vectors ψ and φ.

The solvability condition for equation (3.12) with the reducible-invertible operator Q has the form

$$\Pi(\psi - Q_1\varphi) = 0 \tag{3.13}$$

where Π is the projector to N_Q. Taking into account that $\varphi \in N_Q$, that is $\Pi\varphi = \varphi$, the condition (3.13) leads to the equality

$$\Pi Q_1\Pi\varphi = \Pi\psi. \tag{3.14}$$

Now we come to the decisive step in the analysis of the singular perturbation problem (3.10). The operator $\Pi Q_1\Pi$ acts in the subspace N_Q and $\Pi Q_1\Pi f = 0$, if $f \in R_Q$.

On the contracted space $\hat{N}_Q$ introduce the contracted operator $\hat{Q}_1$ which is determined by the relation

$$\Pi Q_1\Pi = \hat{Q}_1\Pi, \tag{3.15}$$

and also set $\hat{\psi} := \widehat{\Pi\psi} \in \hat{N}_Q$.

Then equality (3.14) on the space $\hat{N}_Q$ has the form

$$\hat{Q}_1\hat{\varphi} = \hat{\psi}. \tag{3.16}$$

The relation (3.16) establishes a connection between two vectors $\hat{\varphi}$ and $\hat{\psi}$ in $\hat{N}_Q$.

The last step is to solve equation (3.12). Using formula (3.9) we get

$$\varphi_1 = R_0(\psi - Q_1\varphi), \quad \Pi\varphi_1 = 0. \tag{3.17}$$

Here R_0 is the potential of the operator Q. At last the vector θ_ϵ has the following representation:

$$\theta_\varepsilon = Q_1\varphi_1 = Q_1 R_0(\psi - Q_1\varphi). \tag{3.18}$$

Equations (3.16)–(3.18) give us the solution of the singular perturbation problem (3.10). With the choice of vectors $\hat{\varphi}$ and $\hat{\psi}$ which satisfy the equality (3.16) we construct vectors φ_1 and θ_ϵ of formulas (3.17) and (3.18).

Consider what follows from the analysis of the equality (3.16). If the operator $\hat{Q}_1$ is invertible, i. e. $\hat{Q}_1^{-1}$ exists, then the equality (3.16) determines one-to-one correspondence between vectors $\hat{\varphi}$ and $\hat{\psi}$.

It means that the problem of singular perturbation (3.10) has a unique solution for an arbitrary vector $\psi \in \mathcal{B}$ under condition $\Pi\varphi_1 = 0$.

Another situation arises when the operator $\hat{Q}_1$ is not invertible. Now there are various possibilities [12]. We consider only two different cases: $\hat{Q}_1 = 0$ is zero operator or, in other terms

$$\Pi Q_1 \Pi = 0. \tag{3.19}$$

This equation is called a *balance condition*. Another case is when the operator $\hat{Q}_1$ is reducible-invertible.

For convenience of references we give the solution of singular perturbation problem (3.10) with invertible operator $\hat{Q}_1$ in the following form.

Lemma 3.2. Let the following condition be satisfied:

(a) bounded operator Q on the Banach space $\mathcal{B}$ is reducible-invertible with projector Π to the null-space $N_Q := \{\varphi : Q\varphi = 0\}$, $\dim N_Q \geq 1$;

(b) the perturbing operator Q_1 on $\mathcal{B}$ is closed with a dense domain $\mathcal{B}_0 \subseteq \mathcal{B}$, $\bar{\mathcal{B}}_0 = \mathcal{B}$;

(c) the contracted operator $\hat{Q}_1$, which is defined by the relation

$$\Pi Q_1 \Pi = \hat{Q}_1 \Pi,$$

has the inverse operator $\hat{Q}_1^{-1}$.

Then the asymptotic representation

$$[\varepsilon^{-1}Q + Q_1](\varphi + \varepsilon\varphi_1) = \psi + \varepsilon\theta_\varepsilon \tag{3.20}$$

is realized by the vectors which are determined by the equality

$$\hat{Q}_1 \hat{\varphi} = \hat{\psi} \qquad (3.21)$$

and relations (3.22) and (3.23):

$$\varphi_1 = R_0(\psi - Q_1\varphi) \qquad (3.22)$$

$$\theta_\varepsilon = Q_1 R_0(\psi - Q_1\varphi). \qquad (3.23)$$

Here $R_0 := [Q + \Pi]^{-1} - \Pi$ is the potential of operator Q.

The proof of Lemma 3.2 was given above in the case of the bounded operator Q_1. In the case of the closed densely defined operator Q_1 Lemma 3.2 is valid with respect to vectors from the domain $\mathcal{B}_0$ of operator Q_1.

Remark 3.1. Lemma 3.2 is valid for the closed densely defined operator Q with common domain $\mathcal{B}_0$ of operators Q and Q_1 [12].

Now we consider the *problem of singular perturbation under the balance condition* (3.19).

Lemma 3.3. Let Q be a bounded reducible-invertible operator on the Banach space $\mathcal{B}$ with the projector Π and the potential R_0. Assume that operators Q_1 and Q_2 are closed with the common domain $\mathcal{B}_0$ and the operator $Q_0 := Q_2 - Q_1 R_0 Q_1$ after contraction on the space $\hat{N}_Q$:

$$\hat{Q}_0\Pi = \Pi Q_0\Pi, \qquad (3.24)$$

has the inverse operator $\hat{Q}_0^{-1}$. In addition, the operator Q_1 satisfies the balance condition

$$\Pi Q_1 \Pi \varphi = 0, \quad \varphi \in \mathcal{B}_0. \qquad (3.25)$$

Then the asymptotic representation

$$[\varepsilon^{-2}Q + \varepsilon^{-1}Q_1 + Q_2](\varphi + \varepsilon\varphi_1 + \varepsilon^2\varphi_2) = \psi + \varepsilon\theta_\varepsilon \qquad (3.26)$$

is realized by the vectors which are determined by the equality

$$\hat{Q}_0 \hat{\varphi} = \hat{\psi} \qquad (3.27)$$

and the relations

$$\varphi_1 = -R_0 Q_1 \varphi; \tag{3.28}$$

$$\varphi_2 = R_0(\psi - Q_0 \varphi); \tag{3.29}$$

$$\theta_3 = [Q_1 + \varepsilon Q_2]\varphi_2 + Q_2 \varphi_1. \tag{3.30}$$

Proof of Lemma 3.3. Expanding the left-hand side of (3.26) in degrees of parameter ε and comparing the results with right-hand side of (3.26), we get the following relations:

$$Q\varphi = 0,$$

$$Q\varphi_1 + Q_1\varphi = 0,$$

$$Q\varphi_2 + Q_1\varphi_1 + Q_2\varphi = \psi,$$

$$[Q_1 + \varepsilon Q_2]\varphi_2 + Q_2\varphi_1 = \theta_\varepsilon. \tag{3.31}$$

The first equality in (3.31) means that $\varphi \in N_Q$, i.e. $\Pi\varphi = \varphi$. Hence, the balance (3.25) yields the solvability of the second equation of (3.31) with respect to the vector φ_1 which can be represented in the form (3.28) under condition $\Pi\varphi_1 = 0$. Substituting (3.28) into the third relation of (3.31), we get an equation with respect to vector φ_2

$$Q\varphi_2 = \psi - Q_0\varphi. \tag{3.32}$$

We can obtain the equality (3.27) from the solvability condition of this equation because $\Pi(\psi - Q_0\varphi) = 0$ is equivalent to $\Pi\psi = \Pi Q_0 \Pi\varphi$, whose contraction on the subspace $\hat{N}_Q$ is (3.27).

The solution of the equation (3.32) under condition $\Pi\varphi_2 = 0$ is represented as (3.29).

Finally the formula (3.30) is obvious.

Remark 3.2. Lemma 3.3 is valid for the closed operator Q with the common dense domain $\mathcal{B}_0$. In this case the potential R_0 is a closed densely defined operator [12].

There are various situations when the operator $\hat{Q}_0$ is not invertible. We will utilize the *situation when the operator $\hat{Q}_1$ is reducible-invertible*.

Lemma 3.4. Suppose that conditions (a) and (b) of Lemma 3.2 are satisfied coincidentally with

(c′) the contracted operator $\hat{Q}_1$ is reducible-invertible with the null-space $\hat{N}_{\hat{Q}_1} \subset \hat{N}_Q$ and with the projector $\hat{\Pi}$ on $\hat{N}_{\hat{Q}_1}$.

(d) The operator Q_2 is closed with the common dense domain $\mathcal{B}_0$ and the twice contracted operator $\hat{\hat{Q}}_2$ on $\hat{N}_{\hat{Q}_1}$, which is determined by the relations

$$\hat{\Pi}\hat{Q}_2\hat{\Pi} = \hat{\hat{Q}}_2\hat{\Pi}, \quad \hat{Q}_2\Pi = \Pi Q_2\Pi, \tag{3.33}$$

has the inverse operator $\hat{\hat{Q}}_2^{-1}$.

Then the asymptotic representation

$$[\varepsilon^{-2}Q + \varepsilon^{-1}Q_1 + Q_2](\varphi + \varepsilon\varphi_1 + \varepsilon^2\varphi_2) = \psi + \varepsilon\theta_\varepsilon \tag{3.34}$$

is realized by the vector in $\mathcal{B}_0$ which are determined by the equation

$$\hat{\hat{Q}}_2\hat{\hat{\varphi}} = \hat{\hat{\psi}} \tag{3.35}$$

and the relations

$$\hat{\varphi}_1 = \hat{R}_0(\hat{\psi} - \hat{Q}_2\hat{\varphi}), \tag{3.36}$$

$$\varphi_2 = R_0(\psi - Q_2\varphi - Q_1\varphi_1), \tag{3.37}$$

$$\theta_\varepsilon = [Q_1 + \varepsilon Q_2]\varphi_2 + Q_2\varphi_1, \tag{3.38}$$

where $\hat{R}_0 := [\hat{Q}_1 + \hat{\Pi}]^{-1} - \hat{\Pi}$ is the potential of the operator $\hat{Q}_1$.

Proof of Lemma 3.4. Recalling the proof of Lemma 3.3 and comparing the left- and right-hand side of (3.34), we conclude that the relations (3.31) are valid. The following analysis is a somewhat different one. The first relation of (3.31) gives us $\Pi\varphi = 0$. Hence the solvability condition for the second equation of (3.31) with respect to the vector φ_1 has the following form:

$$\Pi Q_1\Pi\varphi = 0.$$

After contraction on the subspace $\hat{N}_Q$ we get

$$\hat{Q}_1\hat{\varphi} = 0. \tag{3.39}$$

It means that $\hat{\varphi} \in N_{\hat{Q}_1}$, i. e. $\hat{\Pi}\hat{\varphi} = \hat{\varphi}$.

Then the second relation of (3.31) becomes $Q\varphi_1 = 0$, i. e. $\varphi_1 \in N_Q$.

Now we will construct equations for vectors φ_1 and φ_2. The solvability condition of the third equation of (3.31) with respect to vector φ_2 has the following form:

$$\Pi Q_1 \Pi \varphi_1 + \Pi Q_2 \Pi \varphi = \Pi \psi$$

or we can write it in contracted form

$$\hat{Q}_1 \hat{\varphi}_1 + \hat{Q}_2 \hat{\varphi} = \hat{\psi}. \tag{3.40}$$

Under condition (c') the solvability condition of the equation (3.40) has the form

$$\hat{\Pi}(\hat{\psi} - \hat{Q}_2\hat{\varphi}) = 0$$

or, after contraction to the subspace $\hat{N}_{\hat{Q}_1}$, the form (3.35). Under assumption (d) relation (3.35) establishes one-to-one correspondence between vectors φ and ψ. Now the solution $\hat{\varphi}_1$ of the equation (3.40) under condition $\hat{\Pi}\hat{\varphi} = 0$ is given by (3.36). Formula (3.37) gives the solution of the third equation of (3.31). The solvability condition in this case is equation (3.40). The last relation of (3.31) was explained above.

Let us consider in more detail the *main step of the solution of the singular perturbation problem* in the case of *perturbations of the generator Q of the uniformly ergodic Markov jump process $\kappa(t)$, $t \geq 0$ on a standard phase space $(E, \mathcal{E})$.*

Let us assume that the phase space E *is splitting* into the finite ergodic classes

$$E = \cup_{k=1}^{N} E_k, \quad E_k \cap E_{k'} = \emptyset, \quad k \neq k',$$

with the stationary distributions $\pi_k(dx)$, $1 \leq k \leq N$.

The projector Π into the null-space N_Q of the generator Q acts in the following way:

$$\Pi\varphi(x) = \sum_{k=1}^{N} \hat{\varphi}_k 1_k(x), \quad \hat{\varphi}_k := \int_{E_k} \pi_k(dx)\varphi(x), \tag{3.41}$$

where

$$1_k(x) := \begin{cases} 1, & x \in E_k, \\ 0, & x \notin E_k. \end{cases}$$

Consequently the contracted space $\hat{N}_Q$ is a N-dimensional Euclidean space R^N.

The contracted vector $\hat{\varphi} := \widehat{\Pi\varphi}$ is determined by the components $\hat{\varphi}_k$, $1 \le k \le N$:

$$\hat{\varphi} := (\hat{\varphi}_k; \ 1 \le k \le N).$$

Let Q_1 be a bounded linear operator on the Banach space $\mathcal{B}_E$ of real-valued bounded functions on $\mathcal{E}$ which acts in the following way:

$$Q_1\varphi(x) := \int_E Q_1(x, dy)\varphi(y), \tag{3.42}$$

where the kernel $Q_1(x, dy)$ satisfies the next conditions:

$$Q_1(x, E) = q_1(x), \quad \|q_1(x)\| := \sup_{x \in E} |q_1(x)| < +\infty.$$

Let us calculate the contracted operator $\hat{Q}_1$ which is determined by the relation

$$\hat{Q}_1\Pi = \Pi Q_1\Pi. \tag{3.43}$$

Taking into consideration (3.41) and (3.42) we get

$$\Pi Q_1\Pi\varphi(x) = \Pi Q_1 \sum_{k=1}^N \hat{\varphi}_k 1_k\,(x) = \sum_{k=1}^N \hat{\varphi}_k \Pi \int_{E_k} Q(x, dy)$$

$$= \sum_{k=1}^N \hat{\varphi}_k \Pi Q_1(x, E_k) = \sum_{k=1}^N \hat{\varphi}_k \sum_{r=1}^N \hat{q}_{rk} 1_r\,(x), \tag{3.44}$$

where by definition

$$\hat{q}_{rk} := \int_{E_r} \pi_r(dx)Q_1(x, E_k), \ 1 \le r, k \le N. \tag{3.45}$$

Therefore we rearrange sums in (3.44) and obtain

$$\Pi Q_1\Pi\varphi(x) = \sum_{r=1}^N 1_r\,(x) \sum_{k=1}^N \hat{q}_{rk}\hat{\varphi}_k.$$

According to (3.43) we can conclude that the contracted operator $\hat{Q}_1$ is determined by the matrix

$$\hat{Q}_1 := [\hat{q}_{rk}; \ 1 \le r, k \le N]$$

and acts on the N-dimensional Euclidean space of vectors in the following manner:

$$\hat{Q}_1\hat{\varphi} = \hat{\psi}, \quad \hat{\psi}_r := \sum_{k=1}^{N} \hat{q}_{rk}\hat{\varphi}_k.$$

So the contracted operator $\hat{Q}_1$ is a matrix operator on the contracted N-dimensional Euclidean space $R^N = \hat{N}_Q$.

3.3 Martingale characterization of Markov processes

The concept of the martingale is useful for the modern theory of stochastic processes, in particular for limit theorems for stochastic processes in a series scheme [4, 7, 13].

Let $(\Omega, \mathcal{F}, \mathcal{F}_t, \mathcal{P})$ be a basic probability space with a *filtration* of σ-algebras $\mathcal{F}_t$, $t \in T$. Suppose that the *following conditions* are satisfied:

$$\mathcal{F}_s \subseteq \mathcal{F}_t \text{ if } s < t, \quad \mathcal{F}_t \subseteq \mathcal{F} \text{ for all } t \in T.$$

The set T is finite or infinite interval of positive real numbers: $T := [0; T]$, $T \leq +\infty$. We always assume that the filtration $\mathcal{F}_t, t \in T$ is *right continuous:* $\cap_{s>t}\mathcal{F}_s = \mathcal{F}_{t+} = \mathcal{F}_t$, and is *complete:* $\{A : \mathcal{P}(A) = 0\} \subset \mathcal{F}_0$. The σ-algebras $\mathcal{F}_t$ can be interpreted as the totality of observed events up to the time t.

A stochastic process $\kappa(t)$, $t \in T$ on a measurable phase state space $(E, \mathcal{E})$ is called *adaptive to the filtration* $\mathcal{F}_t$, $t \in T$, if $\kappa(t)$ is $\mathcal{F}_t$-*measurable* for each $t \in T : \{\kappa(t) \in B\} \in \mathcal{F}_t$, $B \in \mathcal{E}$.

For a process $\kappa(t)$ we define $\mathcal{F}_t^{\kappa} := \sigma\{\kappa(s); \ s \leq t\}$, that is σ-algebra of events obtained by the observation of $\kappa(s)$ up to the time t. Since $\mathcal{F}_t$ is increasing in t then $\kappa(t)$ is $\mathcal{F}_t$-adapted if $\mathcal{F}_t^{\kappa} \subset \mathcal{F}_t$ for each $t > 0$.

Definition 3.4. A real-valued process $\mu_t, t \in T$ with $E|\mu_t| < \infty$ for all $t \in T$ adapted to the filtration $\mathcal{F}_t$, $t \in T$, is called $\mathcal{F}_t$-*martingale* (*submartingale*) if

$$E[\mu_t/\mathcal{F}_s] = \mu_s, \quad (E[\mu_t/\mathcal{F}_s] \geq \mu_s) \quad s \leq t, \tag{3.46}$$

with probability 1.

Note that if $E[\mu_t/\mathcal{F}_s] \leq \mu_s$ then μ_t is called $\mathcal{F}_t$-*supermartingale*.

If $\mathcal{F}_t = \mathcal{F}_t^\kappa$ we will say that μ_t is a martingale (sub-, supermartingale).

Examples of martingales.

3.1. The sum $S_n = \sum_{k=0}^n \alpha_k$, $n \geq 0$ of independent random variables with $E\alpha_k = 0$ is a martingale adapted to filtration

$$\mathcal{F}_n^\alpha := \sigma\{\alpha_k;\ 0 \leq k \leq n\}, \quad n \geq 0.$$

3.2. A stochastic process $\kappa(t), t \in T$, with independent increments and $E\kappa(t) = 0$ is a martingale adapted to filtration

$$\mathcal{F}_t^\kappa := \sigma\{\kappa(s);\ 0 \leq s \leq t\},\ t \in T.$$

3.3. Let η be a real-valued random variable with finite mean: $E|\eta| < +\infty$, and $\mathcal{F}_t,\ t \in T$, is given filtration.
   The stochastic process

$$\mu_t := E[\eta/\mathcal{F}_t], t \in T,$$

is a martingale. Moreover, this martingale is *uniformly integrable:*

$$\lim_{C \to \infty} \sup_{t \in T} E[\mu_t I(|\mu_t| \geq C)] = 0,$$

and

$$\sup_{t \in T} E|\mu_t| \leq E|\eta|.$$

Definition 3.5. A martingale μ_t, $\mathcal{F}_t$, $t \in T$, is called *square integrable* if

$$\sup_{t \in T} E\mu_t^2 < +\infty.$$

For square integrable martingale μ_t, $\mathcal{F}_t$, $t \in T$, the stochastic process $\mu_t^2, \mathcal{F}_t, t \in T$, is integrable submartingale [17].
   We shall use *Doob–Meyer decomposition* of a square integrable submartingale [17]:

$$\mu_t^2 = \langle \mu \rangle_t + \mu_t', \tag{3.47}$$

where $\mu'_t, \mathcal{F}_t, t \in T$, is a martingale, and the increasing process $\langle \mu \rangle_t$ is called the *square characteristic of martingale* μ_t. Note that $\langle \mu \rangle_t$ is a nonnegative valued process adapted to filtration $\mathcal{F}_t, t \in T$.

Using the martingale property (3.46) we get

$$E\left[(\mu_t - \mu_s)^2/\mathcal{F}_s\right] = E\left[(\mu_t{}^2 - \mu_s{}^2)/\mathcal{F}_s\right].$$

Comparing this equation with the Doob–Meyer decomposition (3.47) we obtain

$$E\left[(\mu_t - \mu_s)^2/\mathcal{F}_s\right] = E[(\langle \mu \rangle_t - \langle \mu \rangle_s)/\mathcal{F}_s]. \tag{3.48}$$

This formula can be taken as definition of square characteristic of martingale: if an increasing process $\langle \mu \rangle_t, t \in T$, satisfies equality (3.48) for given martingale $\mu_t, \mathcal{F}_t, t \in T$, then this process $\langle \mu \rangle_t, t \in T$, is square characteristic of martingale μ_t.

Martingale induced by a Markov process. Let $\kappa(t), t \geq 0$, be a homogeneous stochastically continuous Markov process defined on basic probability space $(\Omega, \mathcal{F}, \mathcal{F}_t, \mathcal{P})$ with values in a separable metric space $(E, \mathcal{E})$ given by the transition probabilities

$$P_t(x, B) := \mathcal{P}\{\kappa(t) \in B/\kappa(0) = x\}, x \in E, B \in \mathcal{E},$$

and satisfy the Chapman–Kolmogorov equation

$$P_{t+s}(x, B) = \int_E P_t(x, dy) P_s(y, B).$$

On the Banach space $\mathcal{B}_E$ of measurable real-valued functions $\varphi(x), x \in E$, with sup-norm $\|\varphi\| := \sup_{x \in E} |\varphi(x)|$ we can introduce the strongly continuous contraction semigroup of operators $P_t, t \geq 0$, on $\mathcal{B}_E$ by the transition probabilities in the following way:

$$P_t\varphi(x) := \int_E P_t(x, dy)\varphi(y), \quad t \geq 0, \quad \varphi \in \mathcal{B}_E,$$

or using the conditional expectation

$$E_x\varphi(\kappa(t)) := E[\varphi(\kappa(t))/\kappa(0) = x],$$

we can define

$$P_t\varphi(x) := E_x\varphi(\kappa(t)).$$

Let Q be the generator of the semigroup $P_t, t \geq 0$, with the dense domain of definition $\mathcal{B}_0 \subseteq \mathcal{B}$. The semigroup P_t, $t \geq 0$, satisfies the integral equation (cf [14])

$$P_t\varphi(x) = \varphi(x) + \int_0^t QP_s\varphi(x)ds,$$

or using the conditional expectation this equation can be rewritten in the following form:

$$E_x \left[\varphi(\kappa(t)) - \varphi(x) - \int_0^t Q\varphi(\kappa(s))ds \right] = 0.$$

Let $\mathcal{F}_t^\kappa := \sigma\{\kappa(s); \ 0 \leq s \leq t\}$ be the filtration of σ-algebras generated by the Markov process $\kappa(t)$, $t \geq 0$. Then the process

$$\mu_t := \varphi(\kappa(t)) - \int_0^t Q\varphi(\kappa(s))ds, t \geq 0, \tag{3.49}$$

is $\mathcal{F}_t^\kappa$-martingale. Indeed, using the homogeneity and Markovian property of the process $\kappa(t), t \geq 0$, we get for $h > 0$

$$E[(\mu_{t+h} - \mu_t)/\mathcal{F}_t] = E \left[(\varphi(\kappa(t+h)) - \varphi(\kappa(t)) - \int_t^{t+h} Q\varphi(\kappa(s))ds)/\mathcal{F}_t \right]$$

$$= E_{\kappa(0)} \left[\varphi(\kappa(h)) - \varphi(\kappa(0)) - \int_0^h Q\varphi(\kappa(s))ds \right] = 0.$$

The martingale characterization of Markov processes is given by the following theorem.

Theorem 3.1. [4] Let $\kappa(t)$, $t \geq 0$, be a stochastic process on a separable metric space $(E, \mathcal{E})$ adapted to filtration $\mathcal{F}_t, t \geq 0$, and let Q be a generator of strongly continuous contraction semi-group $P_t, t \geq 0$, on the Banach space $\mathcal{B}_E$ with dense domain $\mathcal{B}_0 \subseteq \mathcal{B}_E$. If for all $\varphi \in \mathcal{B}_0$ the process $\mu_t, t \geq 0$, which is given by the relation (3.49) is $\mathcal{F}_t$-martingale then $\kappa(t)$ is Markov process generated by the infinitesimal operator Q.

The *process $\kappa(t)$ is said to solve the martingale problem for the generator* Q.

The next theorem gives the square characteristic of martingale (3.49).

Theorem 3.2. [20] The square characteristic of martingale (3.49) is given by the relation

$$\langle \mu \rangle_t = \int_0^t \left[Q\varphi^2(\kappa(s)) - 2\varphi(\kappa(s))Q\varphi(\kappa(s)) \right] ds. \tag{3.50}$$

Proof of Theorem 3.2. Using the representation (3.49) we calculate

$$\varphi^2 = [\mu + J]^2 = \mu^2 + 2\mu J + J^2 =: \mu^2 + \psi,$$

where

$$\psi = 2\mu J + J^2, \quad J := \int_0^t Q\varphi(\kappa(s))ds.$$

Performing the differentiation of ψ we get

$$d\psi = 2d\mu J + 2\mu Q\varphi ds + 2Q\varphi ds J.$$

Application of (3.49) yields

$$d\psi = 2d\mu J + 2\varphi Q\varphi ds.$$

Now we integrate the last relation

$$\psi = 2\int_0^t d\mu_s \int_0^s Q\varphi dv + 2\int_0^t \varphi Q\varphi ds.$$

The first term is a martingale as integral over martingale μ_s. It is obvious that

$$\varphi^2 - \int Q\varphi^2 ds = \mu'$$

is a martingale too. Hence

$$
\begin{aligned}
\mu^2 &= \varphi^2 - \psi \\
&= \varphi^2 - \int_0^t Q\varphi^2 ds + \int_0^t Q\varphi^2 ds - 2\int_0^t \varphi Q\varphi ds - \mu' \\
&= \int_0^t [Q\varphi^2 - 2\varphi Q\varphi] ds + \mu'',
\end{aligned}
$$

where μ' and μ'' are martingales.

The last relation gives us square characteristic of martingale (3.49) in the form (3.50).

Martingale characterization of Markov chain. Let κ_n, $n \geq 0$, be a Markov chain on a measurable phase space $(E, \mathcal{E})$ induced by a stochastic kernel $P(x, B)$, $x \in E$, $B \in \mathcal{E}$.

Introduce the operator P on the Banach space $\mathcal{B}_E$ of all measurable bounded functions with sup-norm

$$P\varphi(x) := \int_E P(x, dy)\varphi(y) = E[\varphi(\kappa_n)/\kappa_{n-1} = x]. \tag{3.51}$$

Let $\mathcal{F}_n^\kappa := \sigma\{\kappa_k; \ 0 \leq k \leq n\}, n \geq 0$,

be the natural filtration generated by trajectories of Markov chain. The Markov property can be represented in the following form:

$$E[\varphi(\kappa_n)/\mathcal{F}_{n-1}^\kappa] = P\varphi(\kappa_{n-1}). \tag{3.52}$$

Indeed according to the Markov property and definition (3.51)

$$E[\varphi(\kappa_n)/\mathcal{F}_{n-1}^\kappa] = E[\varphi(\kappa_n)/\kappa_{n-1}] = P\varphi(\kappa_{n-1}).$$

Now we construct the martingale as a sum of *martingale differences*

$$\mu_n = \sum_{k=1}^n \left[\varphi(\kappa_k) - E[\varphi(\kappa_k)/\mathcal{F}_{k-1}^\kappa]\right]. \tag{3.53}$$

With the application of Markov property (3.52) and after the rearrangement of the terms the martingale (3.53) yields the form

$$\mu_n = \varphi(\kappa_n) - \varphi(\kappa_0) - \sum_{k=1}^{n-1}[P - I]\varphi(\kappa_k). \tag{3.54}$$

This representation of the martingale is associated with a Markov chain possession of a characterization property.

Lemma 3.5. [4] Let κ_n, $n \geq 0$, be a sequence of random variables taking values in a measurable phase space $(E, \mathcal{E})$ and adapted to filtration of σ-algebras $\mathcal{F}_n$, $n \geq 0$, and let P be a bounded linear positive operator on the Banach space $\mathcal{B}_E$ induced by the transition probabilities $P(x, B)$. If for every

62   Chapter 3

$\varphi(x) \in \mathcal{B}_E$ the right-hand side of (3.54) is a martingale μ_n, $\mathcal{F}_n$, $n \geq 0$, then
the sequence κ_n, $n \geq 0$, is the Markov chain with the transition probabilities
$P(x, B)$ induced by the operator P.

Proof of Lemma 3.5. Using (3.54) we calculate

$$
\begin{aligned}
E[\mu_n/\mathcal{F}_{n-1}] &= E[\varphi(\kappa_n)/\mathcal{F}_{n-1}] - \varphi(\kappa_0) - \sum_{k=1}^{n-1}[P - I]\varphi(\kappa_k) \\
&= E[\varphi(\kappa_n)/\mathcal{F}_{n-1}] - P\varphi(\kappa_{n-1}) + \varphi(\kappa_{n-1}) - \varphi(\kappa_0) \\
&\quad - \sum_{k=1}^{n-2}[P - I]\varphi(x_k) \\
&= \mu_{n-1} + E[\varphi(\kappa_n)/\mathcal{F}_{n-1}] - P\varphi(\kappa_{n-1}).
\end{aligned}
$$

The martingale property

$$
E[\mu_n/\mathcal{F}_{n-1}] = \mu_{n-1}
$$

yields to Markov property (3.52).

By the definition of square characteristic of martingale (3.53) it is easy
to check that

$$
\langle \mu \rangle_n = \sum_{k=0}^{n-1}[P\varphi^2(\kappa_k) - (P\varphi(\kappa_k))^2]. \tag{3.55}
$$

3.4   Pattern limit theorem

Phase merging and averaging algorithms for stochastic systems are based on
limit theorems for random processes in the series scheme. The same applies
to the diffusion approximation of fluctuations as well.

The processes arising in applications and being of interest to us now are
Markovian and with values in a locally compact complete separable metric
space E with σ-algebra of measurable sets $\mathcal{E}$. Such a space $(E, \mathcal{E})$ we call a
standard phase space of the stochastic system.

The functional space $\mathcal{D}_E[0, \infty)$ of right continuous functions $\kappa(t) : [0, \infty) \to$
E with left limits., i.e. for each $t \geq 0 \lim_{s \to t+} \kappa(s) = \kappa(t)$, and $\lim_{s \to t-} \kappa(s) =$

$\kappa(t-0)$ exists, is considered as the space of sample paths of stochastic processes. It is well known [4] that for a standard phase space E the space $\mathcal{D}_E[0,\infty)$ is also a complete separable metric space with the Skorokhod topology.

The main type of convergence of stochastic processes is the *weak convergence of finite-dimensional distributions*, that is for the family of stochastic processes $\kappa^\varepsilon(t), \varepsilon > 0$, with the sample paths in $\mathcal{D}_E[0,\infty)$ there is the process $\kappa(t)$ such that

$$\lim_{\varepsilon \to 0} E\varphi(\kappa^\varepsilon(t_1), \ldots, \kappa^\varepsilon(t_N)) = E\varphi(\kappa(t_1), \ldots, \kappa(t_N)). \qquad (3.56)$$

for any $\varphi(x_1, \ldots, x_N) \in C(E^N)$, where $C(E^N)$ is the space of real-valued bounded continuous functions on E^N and for all finite set $\{t_1, \ldots, t_N\} \subset S$, where S is the dense set in $R_+ := [0,\infty)$. This type of convergence we denote by

$$\kappa^\varepsilon(t) \stackrel{S}{\Rightarrow} \kappa(t) \quad \text{as } \varepsilon \to 0. \qquad (3.57)$$

The more general type of convergence of stochastic processes is the *weak convergence of associated measures*, i.e. the probability distributions

$$P_\varepsilon(B) := \mathcal{P}\{\kappa^\varepsilon(\cdot) \in B\}, \quad B \in \mathcal{B}_E[0,\infty),$$

where $\mathcal{B}_E[0,\infty)$ is the Borel σ-algebra on $\mathcal{D}_E[0,\infty)$, that is

$$\lim_{\varepsilon \to 0} Ef(\kappa^\varepsilon(\cdot)) = Ef(\kappa(\cdot)) \qquad (3.58)$$

for all $f \in C(\mathcal{D}_E[0,\infty))$, where $C(\mathcal{D}_E[0,\infty))$ is the space of real-valued bounded continuous functions on $\mathcal{D}_E[0,\infty)$ in Skorokhod topology [4].

The weak convergence of processes and of associated measures are denoted as follows

$$\kappa^\varepsilon \Rightarrow \kappa, \quad \mathcal{P}_\varepsilon \Rightarrow \mathcal{P} \quad \text{as } \varepsilon \to 0. \qquad (3.59)$$

The connection between two types of convergence of stochastic processes is realized by means of *relatively compactness* of a family of stochastic processes or, the same, of a family of associated measures $\{\mathcal{P}_\varepsilon, \ \varepsilon > 0\}$, i.e. there exists the weak convergence of a sequence $\mathcal{P}_{\varepsilon_n}, \varepsilon_n \to 0$.

Theorem 3.3. [4] Let $\kappa^\varepsilon(t), \varepsilon > 0$, and $\kappa(t)$ be processes with sample paths in $\mathcal{D}_E[0,\infty)$:

(a) if $\kappa^\varepsilon \Rightarrow \kappa$ as $\varepsilon \to 0$, then $\kappa^\varepsilon(t) \overset{S}{\Rightarrow} \kappa(t)$ for set $S = \{t \geq 0 : \ \mathcal{P}\{\kappa(t) = \kappa(t-)\} = 1\}$.

(b) if $\{\kappa^\varepsilon, \varepsilon > 0\}$ is relatively compact and there exists a dense set $S \subset R_+ = [0, \infty)$ such that $\kappa^\varepsilon(t) \overset{S}{\Rightarrow} \kappa(t)$, then $\kappa^\varepsilon(t) \Rightarrow \kappa(t)$ as $\varepsilon \to 0$.

To verify the weak convergence of a family of stochastic processes in $\mathcal{D}_E[0, \infty)$ we have to establish the relatively compactness and the weak convergence of finite-dimensional distributions. All these problems for a family of Markov processes in $\mathcal{D}_E[0, \infty)$ can be realized with the help of the martingale characterization of a Markov process.

Limit theorems for stochastic processes are being demonstrated in two steps: the establishment of relative compactness for a family of measures associated with given processes and the verification of uniqueness of the limiting process. In the case of the Markov processes the martingale approach seems to be the most effective one. It is based on martingale characterization of a Markov processes. To prove the relative compactness of corresponding family of measures, we use here the compactness conditions for square integrable martingales. The uniqueness of the limiting measure follows from the existence of the unique solution of the martingale problem [4].

Theorem 3.4. (*Pattern.*) [3, 4] Let $\kappa^\varepsilon(t)$, $\varepsilon > 0$, $t \geq 0$, be a family of stochastic processes with values in the standard phase space $(E, \mathcal{E})$ and adapted to the given filtration of σ-algebras $\mathcal{F}_t^\varepsilon$, $\varepsilon > 0$, $t \geq 0$.

Let $\kappa(t)$, $t \geq 0$ be a homogeneous stochastically continuous Markov process whose transition probabilities are uniquely determined by the generating operator L which is defined on a dense domain $\mathcal{D}_L$ in the space $C(E)$ and in addition assume that $C_0(E)$, i.e. the space of bounded continuous functions with compact support, is contained in the closure of $\mathcal{D}_L$.

The main supposition is the following asymptotic decomposition for all $\varphi \in \mathcal{D}_L$:

$$\varphi(\kappa^\varepsilon(t)) - \int_0^t L\varphi(\kappa^\varepsilon(s))ds = \mu_t^\varepsilon + \psi_t^\varepsilon \qquad (3.60)$$

where μ_t^ε, $\mathcal{F}_t^\varepsilon$, $t \geq 0$, is a family of square integrable martingales with the quadratic characteristic admitting representation in the form

$$\langle \mu^\varepsilon \rangle_t = \int_0^t \zeta^\varepsilon(s)ds \qquad (3.61)$$

with random functions $\zeta^\varepsilon(s)$, $s \geq 0$, satisfying the following condition for every fixed T and some $\delta > 0$

$$E \sup_{0 \leq t \leq T} |\zeta^\varepsilon(s)|^{1+\delta} \leq C < \infty. \tag{3.62}$$

The process ψ_t^ε, $\varepsilon > 0$ are supposed to satisfy the following condition

$$\sup_{\varepsilon \leq \varepsilon_0} E \sup_{0 \leq t \leq T} |\psi_t^\varepsilon| \to 0, \quad \text{as } \varepsilon_0 \to 0. \tag{3.63}$$

Then the weak convergence

$$\kappa^\varepsilon \Rightarrow \kappa, \quad \text{as } \varepsilon \to 0$$

takes place and the limiting process $\kappa(t)$, $t \geq 0$ is characterized by the martingale

$$\varphi(\kappa(t)) - \int_0^t L\varphi(\kappa(s))ds = \mu_t^0. \tag{3.64}$$

In particular, if condition (3.62) is replaced by the next

$$E \sup_{0 \leq t \leq T} |\zeta^\varepsilon(s)| \to 0, \quad \text{as } \varepsilon \to 0, \tag{3.65}$$

then the limiting process $\kappa(t)$ is given by the solution of the deterministic evolutional equation

$$d\varphi(\kappa(t))/dt = L\varphi(\kappa(t)). \tag{3.66}$$

Proof of Theorem 3.4. By virtue of the representation (3.61) and condition (3.62), the quadratic characteristics of the martingales μ_t^ε, $\varepsilon > 0$, are absolutely continuous and bounded in the probability uniformly in $\varepsilon > 0$. Therefore [3], the family $\langle \mu^\varepsilon \rangle_t$, $\varepsilon > 0$, is relatively compact, and gives the sufficient condition for the compactness of the martingales $\mu_t^\varepsilon, \varepsilon > 0$. Next, taking into account condition (3.63) we can conclude that the processes determined by the left-hand side of the equality (3.60) are relatively compact. Since the expression under the sign of the integral in (3.60) are bounded for $\varphi \in \mathcal{D}_L$, the integral term themselves also form a compact family with respect to $\varepsilon > 0$. Hence the family of processes $\varphi(\kappa^\varepsilon(t))$, $\varepsilon > 0$, is relatively compact for each $\varphi \in C_0(E)$. But for the standard phase space E the relative compactness $\varphi(\kappa^\varepsilon(t))$, $\varepsilon > 0$, for $\varphi \in C_0(E) \subseteq \bar{D}_L$ yields the relatively compactness of processes $\kappa^\varepsilon(t)$, $\varepsilon > 0$ [4]. Taking limit in (3.60) over subsequence $\varepsilon_n \to 0$, we get the martingale representation (3.64) of the limiting process $\kappa(t)$.

Uniqueness solution of the martingale problem for the generator L provides the weak convergence $\kappa^\varepsilon \Rightarrow \kappa$ as $\varepsilon \to 0$.

Under condition (3.65) taking limit in (3.60) we get the deterministic evolutional equation (3.66) for the limiting process $\kappa(t)$.

3.5   Ergodic phase merging

The phase merging effect can be observed for an "almost ergodic" Markov process with an *absorbing state* [11]

Let $\kappa^\varepsilon(t)$, $t \geq 0$, be a Markov process on a measurable phase space $(E, \mathcal{E})$ in the series scheme with a small series parameter $\varepsilon > 0$. Assume that the semi-Markov kernel $Q^\varepsilon(x, \mathcal{B}, t)$ of the Markov process $\kappa^\varepsilon(t)$ depends on the series parameter ε in the form

$$Q^\varepsilon(x, B, t) = P^\varepsilon(x, B)(1 - e^{-q(x)t}), \qquad (3.67)$$

$$P^\varepsilon(x, B) = P(x, B) - \varepsilon P_1(x, B), \qquad (3.68)$$

where $P(x, B)$ is a stochastic kernel defining the distribution of *supporting ergodic imbedded Markov chain* $\kappa_n, n \geq 0$, on the phase space $(E_0, \mathcal{E}_0)$, where

$$E = E_0 \cup \{0\}$$

and state 0 is the absorbing state of the Markov process $\kappa^\varepsilon(t)$. It means that the absorbing probabilities of the imbedded Markov chain κ_n^ε, $n \geq 0$, have the following representation

$$P^\varepsilon(x, \{0\}) = 1 - P^\varepsilon(x, E_0) = \varepsilon P_1(x, E_0) =: \varepsilon p(x).$$

Introduce the *absorbing time* of Markov process $\kappa^\varepsilon(t)$:

$$\zeta_x^\varepsilon := \min\{t : \ \kappa^\varepsilon(t) = 0/\kappa_0^\varepsilon = x\}, \ x \in E_0.$$

According to condition (3.68) it is evident that $\zeta_x^\varepsilon \Rightarrow \infty$ as $\varepsilon \to 0$, and the Markov process $\kappa^\varepsilon(t)$ is "almost ergodic" in the subset of states E_0.

Lemma 3.6. The distribution function

$$\Phi^\varepsilon(t, x) = \mathcal{P}\{\varepsilon\zeta_x^\varepsilon > t\}, \ x \in E_0$$

satisfies the evolutional differential equation

$$d\Phi^\varepsilon(t, x)/dt = \varepsilon^{-1}q(x)\left[\int_{E_0} P^\varepsilon(x, dy)\Phi^\varepsilon(t, y) - \Phi^\varepsilon(t, x)\right] \tag{3.69}$$

with the initial condition $\Phi^\varepsilon(0, x) = 1$.

Proof of Lemma 3.6. Let θ_x^ε be a sojourn time of Markov process $\kappa^\varepsilon(t)$ in state $x \in E_0$. Using the evident equality

$$\Phi^\varepsilon(t, x) = \mathcal{P}\{\varepsilon\zeta_t^\varepsilon > t, \ \theta_x^\varepsilon > t/\varepsilon\} + \mathcal{P}\{\varepsilon\zeta_t^\varepsilon > t, \theta_x^\varepsilon \le t/\varepsilon\}. \tag{3.70}$$

The first term in (3.70) has the following representation

$$\mathcal{P}\{\varepsilon\zeta_t^\varepsilon > t, \theta_x^\varepsilon > t/\varepsilon\} = \mathcal{P}\{\theta_x^\varepsilon > t/\varepsilon\}$$
$$= \exp[-q(x)t/\varepsilon].$$

Using the strong Markov property for the second term in (3.70) we get the representation

$$\mathcal{P}\{\varepsilon\zeta_x^\varepsilon > t, \ \theta_x^\varepsilon \le t/\varepsilon\} = \int_0^{t/\varepsilon} q(x)e^{-q(x)s}ds \int_{E_0} P^\varepsilon(x, dy)\Phi^\varepsilon(t - \varepsilon s, y). \tag{3.71}$$

Notice that under condition $\theta_x^\varepsilon = s$ the absorbing time increases with εs. The change of variable $t - \varepsilon s = \varepsilon s'$ gives us

$$\mathcal{P}\{\varepsilon\zeta_x^\varepsilon > t, \theta_x^\varepsilon \le t/\varepsilon\} = q(x)\int_0^{t/\varepsilon} e^{-q(x)(t/\varepsilon - s)}ds \int_{E_0} P^\varepsilon(x, dy)\Phi^\varepsilon(s, y). \tag{3.72}$$

Combining (3.70)–(3.72) we get the integral renewal equation

$$\Phi^\varepsilon(t, x) - q(x)\int_0^{t/\varepsilon} e^{-q(x)(t/\varepsilon - s)}ds \int_{E_0} P^\varepsilon(x, dy)\Phi^\varepsilon(s, y) = e^{-q(x)t/\varepsilon}$$

which is equivalent to the evolutional equation (3.69).

Remark 3.3. Equation (3.69) in the abstract form can be represented as follows:

$$d\Phi^\varepsilon/dt = [\varepsilon^{-1}Q - Q_1]\Phi^\varepsilon, \quad \Phi^\varepsilon(0, x) = 1, \qquad (3.73)$$

where Q is the generator of the supporting Markov process $\kappa(t)$, $t \geq 0$, which acts in the following way

$$Q\varphi(x) := q(x)\left[\int_{E_0} P(x, dy)\varphi(y) - \varphi(x)\right], \qquad (3.74)$$

and the action of the perturbing operator Q_1 is defined by

$$Q_1\varphi(x) := q(x)\int_{E_0} P_1(x, dy)\varphi(y). \qquad (3.75)$$

The solution of equation (3.73) is connected with the singular perturbation problem considered in Section 3.2.

Note that according to the supposition the supporting Markov process $\kappa(t)$ with the generator Q has the stationary distribution $\pi(dx)$ on E_0 which is connected with the stationary distribution of the imbedded Markov chain $\kappa_n, n \geq 0$, in the following form:

$$q(x)\pi(dx) = q\rho(dx), \quad q := \int_{E_0} q(x)\pi(dx).$$

Let us consider the Laplace transform in variable t of the distribution function $\Phi^\varepsilon(t, x)$

$$\varphi^\varepsilon(x, \lambda) := \int_0^\infty e^{-\lambda t}\Phi^\varepsilon(t, x)dt, \ \text{Re } \lambda \geq 0.$$

Equation (3.73) is reduced to the following one:

$$[\varepsilon^{-1}Q - (Q_1 + \lambda)]\varphi^\varepsilon(x, \lambda) = -1. \qquad (3.76)$$

Theorem 3.5. Let the supporting Markov chain $\kappa_n, n \geq 0$, on the phase space $(E_0, \mathcal{E}_0)$ with the stochastic kernel $P(x, B)$ be uniformly ergodic with the stationary distribution $\rho(B)$, $B \in \mathcal{E}_0$. Then the following limiting result holds:

$$\lim_{\varepsilon \to 0} \varphi^\varepsilon(x, \lambda) = [\hat{q} + \lambda]^{-1} \qquad (3.77)$$

where $\hat{q} = q\hat{p}$, $\hat{p} := \int_{E_0} \rho(dx)p(x)$, $p(x) := P_1(x, E_0)$.

Proof of Theorem 3.5. The operator Q is reducible-invertible as the generator of the uniformly ergodic Markov process $\kappa(t), t \geq 0$. Hence, the singular perturbation problem for the left-hand side of the equation (3.76) by Lemma 3.2 has the following asymptotic representation

$$[\varepsilon^{-1}Q - (Q_1 + \lambda)]\varphi^{\varepsilon} = -1 + \varepsilon\theta_{\varepsilon}$$

where $\varphi^{\varepsilon}(x, \lambda) = \varphi(\lambda) + \varepsilon\varphi_1(x, \lambda)$ and the functions $\varphi(\lambda), \varphi_1(x, \lambda)$ are determined by the relations

$$(\hat{Q}_1 + \lambda)\varphi(\lambda) = 1,$$

$$\varphi_1(x, \lambda) = -R_0 Q_1 \varphi(\lambda),$$

$$\Pi\varphi_1 = 0.$$

The contracted operator $\hat{Q}_1$ is determined by the relation $\Pi Q_1 \Pi = \hat{Q}_1 \Pi$ where Π is the projector to the null-space N_Q of the operator Q which acts in the form

$$\Pi\varphi := \int_{E_0} \Pi(dx)\varphi(x) = \hat{\varphi}1(x).$$

Now we calculate the contracted operator $\hat{Q}_1$

$$
\begin{aligned}
\Pi Q_1 \Pi \varphi(x) &= \Pi Q_1 \hat{\varphi}1(x) \\
&= \hat{\varphi}\Pi q(x)p(x) \\
&= \hat{\varphi}\int_{E_0} \pi(dx)q(x)p(x) \\
&= \hat{\varphi}q\int_{E_0} \rho(dx)p(x) = \hat{\varphi}q\hat{p} \\
&= \hat{\varphi}\hat{q}.
\end{aligned}
$$

It means that the operator $\hat{Q}_1$ acts as the multiplying operator on value $\hat{q} = q\hat{p}$:

$$\hat{Q}_1 \hat{\varphi} := \hat{q}\hat{\varphi}.$$

Taking into account boundedness of vectors $\varphi_1(x, \lambda)$ and $\theta_\varepsilon(x, \lambda)$ we get the limiting result of Theorem 3.5.

The relative weak compactness of the family of distribution functions $\Phi^\varepsilon(t, x)$, $\varepsilon > 0$, and uniqueness of the Laplace transform give us the following result

Corollary 3.1. Under condition of Theorem 3.5 there is the following limiting result:

$$\lim_{\varepsilon \to \infty} \mathcal{P}\left\{\varepsilon \zeta_x^\varepsilon > t\right\} = e^{-\hat{q}t}, \tag{3.78}$$

$$\hat{q} = q\hat{p}.$$

Notice that

$$[\hat{q} + \lambda]^{-1} = \int_0^\infty e^{-\lambda t - \hat{q}t} dt.$$

The phase merging effect of limiting result (3.78) can be formulated as the *phase merging algorithm of stochastic systems with stopping*. Let the stochastic system be described by the a Markov jump process $\kappa(t)$ with stoppings on a phase space $(E, \mathcal{E})$ where $E = E_0 \cup \{0\}$, 0 is the absorbing state of $\kappa(t)$. Let $P_0(x, B)$ be a stochastic kernel on $(E_0, \mathcal{E}_0)$ which is determined by the supported uniformly ergodic Markov chain κ_n^0, $n \geq 0$, with the stationary distribution $\rho(dx)$. Let the stationary probability of stopping for one step

$$\hat{p} := \int_{E_0} \rho(dx) P(x, \{0\}) > 0 \tag{3.79}$$

be small enough.

Then the *stochastic system with stopping is merging to the merged system with two phase states: 1 corresponds to the set E_0 and 0 is the absorbing state*. The merged system is described by the primitive renewal process with stoppage probability $\hat{p}$ and the renewal time between two renewal moments θ has the exponential distribution with the intensity

$$q = \int_{E_0} \rho(dx) q(x). \tag{3.80}$$

According to the stoppage rule for a renewal process the general renewal time before stoppage τ can be represented in the following form

$$\tau = \sum_{k=1}^{\nu} \theta_k, \tag{3.81}$$

where θ_k, $k \geq 1$, are independent and identically exponential distributed random variables with intensity q, and ν is geometrically distributed random variable with parameter $\hat{p}$:

$$\mathcal{P}\{\nu = n\} = \hat{p}(1 - \hat{p})^{n-1}, n \geq 1. \qquad (3.82)$$

Random variables ν and θ_k, $k \geq 1$ are mutually independent.

With our conditions under consideration the general renewal stopping time τ has the exponential distribution with intensity $\hat{q} = \hat{p}q$.

Remark 3.4. The merging algorithm, described before, can be applied to a stochastic system whose evolution is determined by the Markov renewal process $(\kappa_n, \theta_n; n \geq 0)$ with arbitrary semi-Markov kernel

$$Q(x, B, t) = \mathcal{P}\{\kappa_{n+1} \in B, \theta_{n+1} \leq t / \kappa_n = x\} = P(x, B)G_x(t). \qquad (3.83)$$

In this case the formula (3.80) has another interpretation [11], namely

$$q = 1/m,$$

$$m := \int_{E_0} \rho(dx)m(x),$$

$$\begin{aligned} m(x) &= E\theta_x \\ &= \int_0^\infty \bar{G}_x(t)dt, \quad x \in E_0. \end{aligned} \qquad (3.5.84)$$

3.6 Splitting phase merging

Let $\kappa^\varepsilon(t), t \geq 0, \varepsilon > 0$, be a family of Markov jump processes on a standard phase space $(E, \mathcal{E})$ with the generator

$$Q^\varepsilon \varphi(x) = q(x)[\int_E P^\varepsilon(x, dy)\varphi(y) - \varphi(x)]. \qquad (3.85)$$

The transition probabilities $P^\varepsilon(x, dy)$ of MJC κ_n^ε, $n \geq 0$, depend on a small series parameter $\varepsilon > 0$ in the form

$$P^\varepsilon(x, dy) = P(x, B) + \varepsilon P_1(x, B). \qquad (3.86)$$

The stochastic kernel $P(x, B)$ is determined as the supporting imbedded Markov chain κ_n, $n \geq 0$, above, on the *splitting phase space*

$$E = \cup_{k=1}^{N} E_k, \quad E_k \cap E_{k'} = \emptyset, \quad k \neq k'. \tag{3.87}$$

The stochastic kernel $P(x, B)$ is coordinated with the splitting (3.87) in the following way:

$$P(x, E_k) = 1_k(x) := \begin{cases} 1, & x \in E_k, \\ 0, & x \notin E_k. \end{cases} \tag{3.88}$$

The main assumption is that the supporting MJC κ_n, $n \geq 0$, is *uniformly ergodic* on every class E_k, $1 \leq k \leq N$, with the stationary distributions

$$\rho_k(B) = \int_{E_k} \rho_k(dx) P(x, B), \quad B \in \mathcal{E}_k, \quad \rho_k(E_k) = 1. \tag{3.89}$$

Here $\mathcal{E}_k := \mathcal{E} \cap E_k$ is contracting of σ-algebra on subset E_k. The supporting Markov process $\kappa(t)$, $t \geq 0$, with the generator

$$Q\varphi(x) = q(x)[\int_E P(x, dy)\varphi(y) - \varphi(x)] \tag{3.90}$$

will be uniformly ergodic as well and will have the stationary distributions $\pi_k(dx), 1 \leq k \leq N$:

$$\pi_k(dx)q(x) = q_k \rho_k(dx), \quad q_k := \int_{E_k} \pi_k(dx)q(x).$$

Introduce the merging function $k(x) = k, x \in E_k$, $1 \leq k \leq N$. The main condition for the *phase merging algorithm* (PMA) is positiveness of the stationary exit probabilities for one step of MJC κ_n^ε, $n \geq 0$, from ergodic classes E_k of the supporting MJC κ_n, $n \geq 0$:

$$\hat{p}_k := \int_{E_k} \rho_k(dx) P_1(x, E \backslash E_k) > 0, \quad 1 \leq k \leq N. \tag{3.91}$$

The result of the splitting phase merging is the following

Theorem 3.6. With assumption given above there is a weak convergence of the splitting processes:

$$k(\kappa^\varepsilon(t/\varepsilon)) \Rightarrow \hat{k}(t), \quad \varepsilon \to 0$$

to the limiting merged Markov process $\hat{\kappa}(t)$, $t \geq 0$, on the merged phase space $\hat{E} = \{1, 2, \ldots, N\}$ with the generating matrix $\hat{Q} = [\hat{q}_{rk}; \ 1 \leq r, k \leq N]$ whose elements are determined by the relations

$$\hat{q}_{rk} := q_r \hat{p}_{rk}, \quad r \neq k, \quad \hat{q}_r := -q_r \hat{p}_{rr};$$

$$\hat{p}_{rk} := \int_{E_r} \rho_r(dx) P_1(x, E_k), 1 \leq r, k \leq N. \tag{3.92}$$

Proof of Theorem 3.6. At first we will clarify the fact that the matrix $\hat{Q}$ with elements determined by (3.92) is the generating matrix of some Markov jump process. Indeed, using relation (3.86) and property (3.88) of the stochastic kernel $P(x, B)$ we get

$$
\begin{aligned}
\varepsilon \hat{p}_{rk} &= \int_{E_r} \rho_r(dx) \varepsilon P_1(x, E_k) \\
&= \int_{E_r} \rho_r(dx) [P^{\varepsilon}(x, E_k) - P(x, E_k)] \\
&= \int_{E_k} \rho_r(dx) P^{\varepsilon}(x, E_k) - \delta_{rk},
\end{aligned}
$$

where δ_{rk} is the Kronecker symbol:

$$\delta_{rk} = \begin{cases} 1, & r = k, \\ 0, & r \neq k. \end{cases}$$

Hence, $\hat{p}_{kr} \geq 0$ if $r \neq k$ and $\hat{p}_{rr} \leq 0$.

Using the evident relation $P_1(x, E) = 0$, which is a direct consequence of relation (3.86) because $P^{\varepsilon}(x, E) = P(x, E) = 1$ and taking into account relation (3.92) we get the following

$$
\begin{aligned}
\hat{p}_k &= -\int_{E_k} \rho_k(dx) P_1(x, E_k) \\
&= -\hat{p}_{kk} \\
&= \sum_{r \neq k} p_{kr}.
\end{aligned}
$$

After multiplication by q_k the last equality in combination with (3.92) gives

$$\hat{q}_k = \sum_{r \neq k} \hat{q}_{kr}.$$

This is the condition for the *conservatism* of generating matrix $\widehat{Q}$ determining the Markov jump process. Condition (3.91) ensures that all states in $\widehat{E} = \{1, 2, \ldots, N\}$ are stable.

In order to prove Theorem 3.6, we shall use the martingale characterization of Markov processes (Section 3.3) $\kappa_t^\varepsilon := \kappa^\varepsilon(t/\varepsilon)$, $\varepsilon > 0$, with the generators

$$Q^\varepsilon := \varepsilon^{-1}Q + Q_1, \qquad (3.93)$$

where the supporting operator Q is defined in (3.90) and the perturbing operator Q_1 acts in the following form

$$Q_1\varphi(x) = q(x)\int_E P_1(x, dy)\varphi(y). \qquad (3.94)$$

Let us consider the martingale which is generating by the Markov process κ_t^ε:

$$\mu_t^\varepsilon = \varphi^\varepsilon(\kappa_t^\varepsilon) - \int_0^t Q^\varepsilon\varphi^\varepsilon(\kappa_s^\varepsilon)ds. \qquad (3.95)$$

We just apply the source idea of work [14]. The martingale characterization of Markov processes with the singular perturbed operator (3.93) is considered on the test functions φ^ε which depend on the series parameter ε in the form

$$\varphi^\varepsilon(x) = \hat{\varphi}(k(x)) + \varepsilon\varphi_1(x). \qquad (3.96)$$

Now we investigate the integrand in (3.95)

$$Q^\varepsilon\varphi^\varepsilon = [\varepsilon^{-1}Q + Q_1](\varphi + \varepsilon\varphi_1). \qquad (3.97)$$

So, we get the singular perturbation problem with the supporting operator Q. Q is reducible-invertible operator with projector Π on the null-space N_Q which acts in the form (Section 3.2)

$$\Pi\varphi(x) = \sum_{k=1}^N \hat{\varphi}_k \mathbb{1}_k(x),$$

$$\hat{\varphi}_k := \int_{E_k} \pi_k(dx)\varphi(x).$$

Note, that the dimension $N = \dim N_Q$ is the number of ergodic classes in splitting (3.87).

Application of Lemma 3.2 to the singular perturbation problem (3.97) give us the following asymptotic representation of the integrand

$$Q^\varepsilon \varphi^\varepsilon(x) = \hat{Q}_1 \hat{\varphi}(k(x)) + \varepsilon \theta^\varepsilon(x), \qquad (3.98)$$

where the merged operator $\hat{Q}_1$ is determined by the relation

$$\Pi Q_1 \Pi = \hat{Q}_1 \Pi,$$

and the vector $\theta(x)$ has the following representation

$$\theta(x) = -Q_1 R_0 Q_1 \hat{\varphi}(k(x)).$$

Here $R_0 := [Q + \Pi]^{-1} - \Pi$ is the potential of the generator Q.

Substituting (3.98) into the martingale given by (3.95), we obtain the representation

$$\hat{\varphi}(k(\kappa_t^\varepsilon)) - \int_0^t \hat{Q}_1 \hat{\varphi}(k(\kappa_s^\varepsilon)) ds = \mu_t^\varepsilon + \varepsilon \psi_t^\varepsilon,$$

where

$$\psi_t^\varepsilon := \int_0^t \theta^\varepsilon(\kappa_s^\varepsilon) ds - \varphi_1(\kappa_t^\varepsilon).$$

Now we can use Theorem 3.4 (Section 3.4). It is easy to check that the condition (3.60)–(3.63) of Theorem 3.4 are valid. So, the family of processes $k(\kappa_t^\varepsilon)$, $\varepsilon > 0$, are weakly compact and the limiting process $\hat{\kappa}(t)$ is the solution of the martingale problem

$$\hat{\varphi}(\hat{\kappa}(t)) - \int_0^t \hat{Q}_1 \hat{\varphi}(\hat{\kappa}(s)) ds = \hat{\mu}_t$$

with the generating operator $\hat{Q}_1$ determining the merged Markov process $\hat{\kappa}(t)$ on the merged phase space $\hat{E} = \{1, 2, \dots, N\}$.

The phase merging effect of the limiting result of Theorem 3.6 can be realized as the *phase merging algorithm of stochastic system with splitting phase space*.

Let the stochastic system be described by a Markov renewal process $\tilde{\kappa}_n, \theta_n; n \geq 0$, on a standard phase space $(E, \mathcal{E})$ with the semi-Markov kernel

$$Q(x, B, t) = P\{\kappa_{n+1} \in B, \theta_{n+1} \leq t / \kappa_n = x\}$$

$$= \tilde{P}(x, B)(1 - e^{-q(x)t}). \tag{3.99}$$

Let $P(x, B)$ be a supporting stochastic kernel which is coordinated with splitting phase space (3.87) and defines the uniformly ergodic Markov chain κ_n, $n \geq 0$, with the stationary distributions (3.89).

Introduce the stationary exit probabilities by the relation (compare with (3.91))

$$\hat{p}_k := \int_{E_k} \rho_k(dx)\tilde{P}(E\backslash E_k) > 0, \quad 1 \leq k \leq N. \tag{3.100}$$

Then, under condition that the stationary exit probabilities are small enough, the merged Markov process $\hat{\kappa}(t)$, $t \geq 0$, on the merged phase space $\hat{E} = \{1, 2, \ldots, N\}$ can be considered as the merged mathematical model of the initial stochastic system whose evolution is described by the Markov renewal process $(\kappa_n, \theta_n;\ n \geq 0)$ with the semi-Markov kernel (3.99).

Hence, there is the following approximate relation for some $T > 0$:

$$k(\tilde{\kappa}(Tt)) \simeq \hat{\kappa}(t), \tag{3.101}$$

where by definition $\tilde{\kappa}(t) := \kappa_{\nu(t)}$, $\nu(t) := \max\{n :\ \tau_n \leq t\}$, $\tau_n := \sum_{k=1}^{n} \theta_k$, $n \geq 0$, $\tau_0 = 0$.

The relation (3.101) means that the evolution of the merged stochastic system $k(\tilde{\kappa}(Tt))$ which is obtained from the initial one is approximately as good as the evolution of the merged stochastic system $\hat{\kappa}(t)$.

3.7   Heuristic phase merging principles

Applications of the phase merging algorithms to real stochastic systems require some complicated and tedious calculations for the construction of an adequate mathematical model of the system as well as for clarifying the stationary distribution of the supporting Markov chain and subsequent calculation of the merged characteristics.

However, it is interesting to notice that there exist some heuristic phase merging principles which give the same result without tiresome technical investigations. Such principles can be considered as some 'evolutional laws' for stochastic systems.

3.7.1 Basic assumptions

Heuristic phase merging principles can be applied under some reasonable assumptions.

1. Finiteness. A system consists of a finite number of elements (devices, subsystems, etc.). Every element of the system can be detected in a finite set of states.

2. Independence. Elements of the system are functioning independently from each other. Changes in position of some element don't have any influence on the evolution of other elements.

3. Semi-Markov property. Functioning of every element is described by the Markov renewal process $(\kappa_n^{(k)}, \theta_n^{(k)}; n \geq 0)$ on a finite phase states space $E_k = \{e_1^{(k)}, e_2^{(k)}, \ldots, e_{N_k}^{(k)}\}$ with the semi-Markov matrix $Q^{(k)}(t) = [Q_{ij}^{(k)}(t); i, j \in E_k]$, $1 \leq k \leq N$ where N is the number of elements in the system.

4. Ergodicity. The imbedded Markov chain $(\kappa_n^{(k)}; n \geq 0)$ with the transition matrices $P^{(k)} = [p_{ij}^{(k)} = Q_{ij}^{(k)}(+\infty); i, j \in E_k]$ are ergodic with the stationary distributions $\rho^{(k)} = (\rho_i^{(k)}; i \in E_k)$. Moreover, in the stationary regime the residual (defect) sojourn time in the states have the stationary distributions

$$G_i^{(k)*}(t) := \lambda_i^{(k)} \int_0^t \bar{G}_i^{(k)}(s)ds,$$

where

$$\lambda_i^{(k)} := 1/m_i^{(k)},$$

$$m_i^{(k)} = E\theta_i^{(k)}$$
$$= \int_0^\infty \bar{G}_i^{(k)}(s)ds,$$

$$\bar{G}_i^{(k)}(s) := 1 - G_i^{(k)}(s)$$
$$= \sum_{j \in E_k} Q_{ij}^{(k)}(s).$$

5. Stability. Among the states of every element there is some stoppage state which we will mark as "zero-state": "0". In the zero-state the element loses its working ability, and there is a positive probability of stopping (failure, etc.) of the whole system.

Stability of the system means that the stoppage probabilities of the system in the zero-states of its elements are rather small as compared with renewal probabilities of the elements. A quantitative characteristic of stability of the system is determined by the phase merging algorithm for the given supporting system without zero-states.

3.7.2 Heuristic phase merging principles

It is easy to see that the functioning stochastic system with a finite number of independent semi-Markov elements can't be described exactly by the Markov renewal process on the direct product of phase space $E = \otimes_{k=1}^{N} E_k$. It is necessary to add some continuous components in order to preserve the semi-Markov property of the system states.

But using the ergodic property, we can retain the discrete phase space of the system. For the sake of greater clarity we will consider further *two-component* stochastic systems. Multi-component systems can be dealt with in a suitable way, and will require some additional technical reasoning.

So, let the system consists of two elements, and every element is described by a Markov renewal process $(\kappa_n^{(k)}, \theta_n^{(k)}; n \geq 0)$ in the finite phase states space

$$
\begin{aligned}
E &= \{0, 1, \ldots, N\} \\
&= [Q_{ij}^{(k)}(t); i, j \in E], \quad k = 1, 2.
\end{aligned}
$$

The discrete phase space of the system is determined as follows:

$$E = \{kij; i, j \in E_0, k = 1, 2\} \qquad (3.102)$$

So, k is the number of the element which changes the state, and i, j are the states of the first and the second elements immediately after the state of the k-th element was changed.

Principle of stationary phase merging. The sojourn time in discrete states of a two-component system are determined by the relations:

$$\theta_{1ij} = \theta_i^{(1)} \wedge \theta_j^{(2)*},$$

$$\theta_{2ij} = \theta_i^{(1)*} \wedge \theta_j^{(2)}. \tag{3.103}$$

The random variables $\theta_i^{(k)*}$ have the stationary distribution

$$P\{\theta_i^{(k)*} \leq t\} = G_i^{(k)*}(t) = \lambda_i^{(k)} \int_0^t \bar{G}_i^{(k)}(s)ds \tag{3.104}$$

The stochastic transit event of the system is described by the relations

$$\{1ij \rightarrow 1i'j\} = \{\theta_i^{(1)} < \theta_j^{(2)*}, i \rightarrow i'\},$$

$$\{1ij \rightarrow 2ij'\} = \{\theta_i^{(1)} \geq \theta_j^{(2)*}, j \rightarrow j'\},$$

$$\{2ij \rightarrow 2ij'\} = \{\theta_i^{(1)*} \geq \theta_j^{(2)}, j \rightarrow j'\},$$

$$\{2ij \rightarrow 1i'j\} = \{\theta_i^{(1)*} < \theta_j^{(2)}, i \rightarrow i'\}. \tag{3.105}$$

The stationary distribution of the system's states is determined as follows:

$$\rho_{1ij} = \rho \rho_i^{(1)} \rho_j^{(2)} m_j^{(2)} \quad \rho_{2ij} = \rho \rho_i^{(1)} \rho_j^{(2)} m_i^{(1)} \tag{3.106}$$

$$\rho := [\sum_{i \in E} \rho_i^{(1)} m_i^{(1)} + \sum_{j \in E} \rho_j^{(2)} m_j^{(2)}]^{-1}$$

where $m_i^{(k)} := E\theta_i^{(k)}, \quad i \in E, \ k = 1, 2$.

According to (3.103) in the stationary regime the system is in such a way that after one element has changed its state the sojourn time of the other element has a stationary distribution (3.104).

We assume that every element has the zero-state and the stoppage of the system occurs when both elements are in their zero-states.

Principle of the absence of aftereffects. The common working time of the system to stoppage τ is determined by the exponential distribution

$$P\{\tau > t\} = e^{-\Lambda t}, \ t \geq 0 \tag{3.107}$$

The stoppage intensity Λ is independent of the initial state of the system.

Principle of superposition of stoppages. The intensity of the system stoppage is determined by the sum of intensities of the system stoppages in the zero-states of every element

$$\Lambda = \Lambda_1 + \Lambda_2 \qquad (3.108)$$

Principle of the independence of stoppages. The intensity of the system stoppage in the zero-state of every element is determined by the stoppage rule of a renewal process with thin out (Section 1.2):

$$\Lambda_k = q_k \lambda_k,$$
$$\lambda_k := [\sum_{i \in E} \rho_i^{(k)} m_i^{(k)}]^{-1}, \qquad (3.109)$$

where q_k is the probability of the system stoppage in the zero-state of the k-th element:

$$q_k := \sum_{i \in E} \rho_i^{(k)} p_{i0}^{(k)}. \qquad (3.110)$$

The *phase merging algorithm is determined by the formulas* (3.107)–(3.1 10).

Remark 3.5. The sojourn time in the system states (3.103) is based on the property of the residual renewal time of the renewal process (Section 1.2). But for the two-component system formula (3.103) which is more ge neral because the changing moments of states of elements are random and moreover, we have been dealing with a Markov renewal process.

Remark 3.6. The exponential distribution of the common stoppage time of the system can be described by the following reasoning. Before the stoppage of the system with rather small probabilities of stoppages q_k there are many changings of states and as a consequence, the stoppage time isn't depend on the initial state.

Remark 3.7. The least evident are the principles of the superposition and independence of stoppages. But under consideration of some attempt to the system stoppage there are many changings of states before a new attempt will occur. And what follows is the principle of the absence of aftereffect.

3.7.3 Heuristic phase merging

Here we will consider some applications of the heuristic phase merging principles to two-component renewal systems.

Double renewal system. There are two elements functioning independently of each other. Every element is described by the alternative renewal process (Section 1.2) with the distribution functions of working and repairing times

$$G_i^{(k)}(t) = \mathcal{P}\{\theta_i^{(k)} \leq t\}, \quad i = 0, 1, \ k = 1, 2.$$

Where $\theta_1^{(k)}$ is the working time of the k-the element and $\theta_0^{(k)}$ is the repairing time of the k-th element.

Such a system in the literature is called a "two lifts system".

According to the stationary phase merging the phase space of the system under consideration contains eight states:

$$E = \{kij; \ k = 1, 2, \ i, j = 0, 1\}$$

The states 100 and 200 are the stoppage states of the system (see Fig. 3.1).

The sojourn times in states are determined by formula (3.103). Using formulas (3.103)–(3.10 5), we can construct the semi-Markov matrix which describes the evolution of this system.

But we can use the heuristic phase merging principles in order to investigate the stoppage intensity of the system. According to the superposition and independence stoppage principles, the stoppage intensity of the system has the following representation:

$$\Lambda = q_1 \lambda_1 + q_2 \lambda_2, \tag{3.111}$$

where $\lambda_1 = 1/E\theta_1^{(1)}$, $\lambda_2 = 1/E\theta_2^{(2)}$.

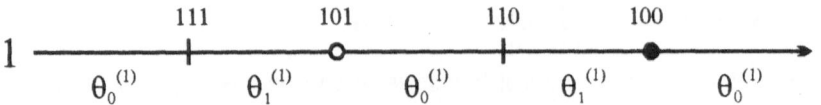

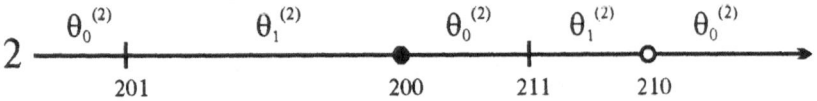

Fig. 3.1. **Double renewal system**

In order to get a formula for the stoppage probabilities q_k, we not e that the stoppage of the system can occur if the working time of one element is finished during the repairing time of the other element. So we get

$$q_1 = \mathcal{P}\{\theta_1^{(2)*} < \theta_0^{(1)}\},$$

$$q_2 = \mathcal{P}\{\theta_1^{(1)*} < \theta_0^{(2)}\} \tag{3.112}$$

Considering the distribution of the stationary sojourn time and independence of random variables $\theta_i^{(k)}$, we obtain the following representation for the stoppage probabilities:

$$q_1 = \lambda_2 \int_0^\infty \bar{G}_0^{(1)}(t)\bar{G}_1^{(2)}(t)\,dt,$$

$$q_2 = \lambda_1 \int_0^\infty \bar{G}_0^{(2)}(t)\bar{G}_1^{(1)}(t)\,dt. \tag{3.113}$$

The stoppage intensity of the double renewal system is determined by formulas (3.111)–(3.113).

Remark 3.8. Notice that for two independent positive random variables α and β with distribution function $\bar{G}_\alpha(t) = \mathcal{P}\{\alpha > t\}$ and $\bar{G}_\beta(t) = \mathcal{P}\{\beta > t\}$

$$\int_0^\infty \bar{G}_\alpha(t)\bar{G}_\beta(t)dt = E(\alpha \wedge \beta).$$

Therefore the stoppage intensity of the double renewal system has the following representation:

$$\Lambda = [E(\theta_0^{(1)} \wedge \theta_1^{(2)}) + E(\theta_1^{(1)} \wedge \theta_0^{(2)})]/E\theta_1^{(1)}E\theta_1^{(2)}. \qquad (3.114)$$

The phase merging principles are acting effectively under the condition of smallness of the stoppage probabilities q_k. This is true when the repairing times $\theta_0^{(k)}$ are essentially smaller than the working times $\theta_1^{(k)}$. So we can assume rough equalities

$$E(\theta_0^{(k)} \wedge \theta_1^{(k')}) \simeq E\theta_0^{(k)}.$$

Then the stoppage intensity of the system can be represented as follows:

$$\Lambda = E(\theta_0^{(1)}) + E(\theta_0^{(2)})/E\theta_1^{(1)}E\theta_1^{(2)}. \qquad (3.115)$$

Introduce the reliability coefficients

$$\delta_k := E\theta_0^{(k)}/E\theta_1^{(k)}.$$

Now the stoppage intensity of the system is determined by the formula

$$\Lambda = \delta_1\lambda_2 + \delta_2\lambda_1. \qquad (3.116)$$

In the particular case of the system with two identical elements we get

$$\Lambda = 2E\theta_0/(E\theta_1)^2. \qquad (3.117)$$

Protective system. There is one working device and one protective facility. The functioning of the working device is described by the simple renewal process with the distribution function of working time $G(t) = \mathcal{P}\{\alpha \leq t\}$.

The operation of the protective facility is described by the alternative renewal process (Section 1.2.) with the
distribution functions $F_k(t) = \mathcal{P}\{\beta_k \leq t\}$, $k = 1, 0$, where β_1 is the action time and β_0 is the renewal time of the protective facility (see Fig. 3.2).

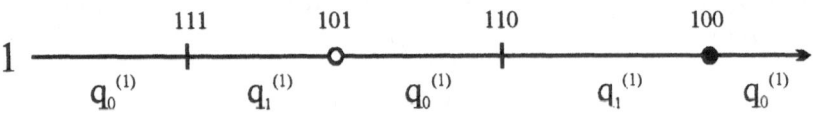

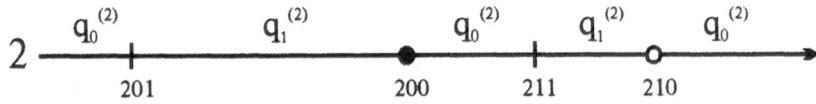

Fig. 3.2. **Protective system**

The system wrecking stoppage occurs when the renewal moment of the wor king device takes place during the renewal time of the protective facility. As usualy, we assume that the stability condition is satisfied, which means $E\beta_0 \ll E\beta_1$.

Using the heuristic principles of phase merging, we will determine the distribution of the stoppage time of the protective system. At first, according to the principle of the absence of aftereffects, we conclude that the stoppage time has the exponential distribution:

$$\mathcal{P}\{\tau > t\} = e^{-\Lambda t}, \ t \geq 0 \tag{3.118}$$

with some intensity Λ which is determined, according to the principle of independence of stoppages, by formula (3.109):

$$\Lambda = q\lambda_1 \quad \lambda_1 = 1/E\beta_1. \tag{3.119}$$

The stoppage probability q in the renewal period of the protective facility is determined, according to the principle of stationary state (cf. with (3.112))

$$q = \mathcal{P}\{\alpha^* < \beta_0\} = \lambda \int_0^\infty \bar{F}_0(t)\bar{G}(t)dt \qquad (3.120)$$

where $\lambda = 1/E\alpha$.

Therefore the stoppage time of the protective system has intensity

$$\Lambda = \int_0^\infty \bar{F}_0(t)\bar{G}(t)dt \Big/ \int_0^\infty \bar{G}(t)dt \int_0^\infty \bar{F}_0(t)dt. \qquad (3.121)$$

Notice that the first integral in (3.121) can be simplified as follows:

$$\int_0^\infty \bar{F}_0(t)\bar{G}(t)dt \simeq \int_0^\infty \bar{F}_0(t)dt.$$

So we get the simple rough expression for the intensity of the stoppage time of the protective system

$$\Lambda = E\beta_0/E\alpha E\beta_1. \qquad (3.122)$$

This formula can be useful in the optimization problem.

Reserved systems. There are two elements functioning independently of each other. Every element is described by the recurrent flow (Section 1.2) with the distribution functions

$$P_i(t) = \mathcal{P}\{\theta_n^{(i)} \le t\}, \quad i = 1, 2, \quad n \ge 1.$$

The renewal times $\theta_n^{(i)}$ are working for all the elements. The replacement of elements is realized instantaneously at every renewal moment:

$$\tau_n^{(i)} = \sum_{k=1}^n \theta_k^{(i)}, \quad n \ge 1, \quad i = 1, 2.$$

That is an ideal reserved system. The real reserved system possesses certain restriction on the changing procedure. Here we assume that the reserving facility works with an initial random time α with a given distribution function

$$G(t) = \mathcal{P}\{\alpha \le t\}.$$

When the renewal time between two adjacent renewal moments is less then the inertial time α, then the real reserved system takes the wrecking stoppage (see Fig. 3.3).

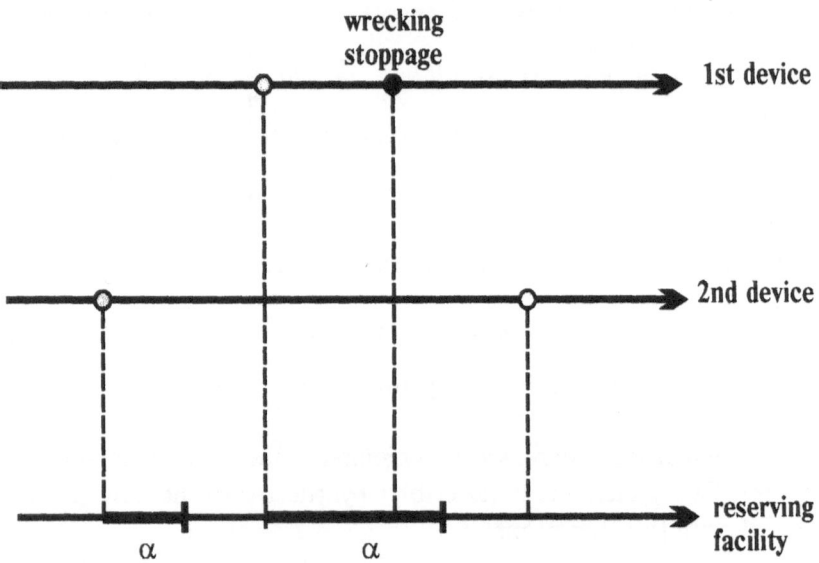

Fig. 3.3. **Reserved system**

According to the superposition procedure considered in Section 1.2.4, the sojourn times of the reserved system can be of two types:

$$\theta_n^{(1)} \wedge x \quad \text{and} \quad \theta_n^2 \wedge x.$$

Using the stationary distribution (1.25) of the Markov chain with the transition probability (1.24) describing the renewal moments of the reserved system, the sojourn times of the stationary regime can be considered as follows (see (3.103)):

$$\theta_n^{(1)} \wedge \theta_n^{(2)*} \quad \text{and} \quad \theta_n^{(1)*} \wedge \theta_n^{(2)}$$

where $\theta_n^{(i)*}$ means the stationary residual time with the distribution density

$$P_i^*(t) = \bar{P}(t)/p_i, \quad p_i := E\theta_n^{(i)} = \int_0^\infty \bar{P}_i(t)dt.$$

Now by the heuristic phase merging principles the stoppage intensity of the real reserved system with an inertial reserved facility has the following representation:

$$\Lambda = q\lambda$$

where the stationary probability of stopping can be calculated by the formula:

$$q = \rho_1 p_{10} + \rho_2 p_{20}.$$

Here

$$\rho_1 = m_2\rho, \quad \rho_2 = m_1\rho, \quad \rho = [m_1 + m_2]^{-1}$$

are the stationary probabilities of the supporting Markov chain $(\delta_n^*,\ n \geq 0)$ with two states $(1, 2)$.

The stopping probabilities p_{i0}, $i = 1, 2$, are determined by the relations

$$p_{10} = \mathcal{P}\{\theta_n^{(1)} \wedge \theta_n^{(2)*} < \gamma\},$$

$$p_{20} = \mathcal{P}\{\theta_n^{(1)*} \wedge \theta_n^{(2)} < \gamma\}.$$

The intensity λ, according to the heuristic principle (3.109) is calculated by the formulas

$$\lambda = [\rho_1 m^{(1)} + \rho_2 m^{(2)}]^{-1},$$

$$m^{(1)} = E\theta_n^{(1)} \wedge \theta_n^{(2)*}, \quad m^{(2)} = E\theta_n^{(1)*} \wedge \theta_n^{(2)}.$$

After some tedious calculations we obtain

$$\lambda = 1/m_1 + 1/m_2.$$

Hence the stopping intensity is represented as follows

$$\Lambda = p_{10}/m_1 + p_{20}/m_2.$$

Chapter 4

Evolutional stochastic system in a random medium

The evolution of systems in a random medium is considered as an interaction of two processes: a *switched process* describes the evolutional system and a *switching process* determines the alternations of the medium.

As the mathematical models of the random medium the *stationary Markov processes* are relevant. The evolution of the system in a random medium is determined by a *random evolution*. The additive functionals on Markov processes and the storage processes are analyzed as the specific interpretation of random evolutions.

Assumptions of *ergodicity* of the random medium and *stationary state* of the evolutional system allow to simplify the description which is given in the form of average limit theorems and diffusion approximation scheme.

4.1   Stochastic additive functionals

Additive functionals on stationary Markov jump processes can be described as the *switched process with stationary independent increments*.

The process with stationary independent increments $\xi(t), t \geq 0$, taking real values, is determined by the following properties [19]:

(a) (*Stationary increments*) The distribution of increments

$$\xi(t_0 + t) - \xi(t_0)$$

depends only on t but not on t_0.

(b) (*Independent increments*) The random variables $\xi(t_{k+1}) - \xi(t_k)$, $1 \le k \le n$, for any finite set of $0 < t_1 < \ldots < t_n$ are mutually independent.

The main properties of the processes with stationary independent increments have *infinite divisibility* of its distribution and *Markov property*.

The characteristic function of increments

$$f_t(\lambda) := E \exp[i\lambda(\xi(t + t_0) - \xi(t_0))]$$

satisfies the semigroup property:

$$f_{t+t'}(\lambda) = f_t(\lambda) f_{t'}(\lambda).$$

Moreover, the characteristic function has the natural representation

$$f_t(\lambda) = \exp[t\Gamma(\lambda)]. \tag{4.1}$$

The *cumulant* $\Gamma(\lambda)$ has the well-known Lévi–Khintchine representation:

$$\Gamma(\lambda) = i\lambda a - \sigma^2\lambda^2/2 + \int_{-\infty}^{+\infty} [e^{i\lambda u} - 1 - i\lambda u I_{(|u|\le 1)}] H(du), \tag{4.2}$$

where the spectral measure $H(du)$ satisfies the following conditions:

$$\int_{|u|\le 1} u^2 H(du) < \infty,$$

$$\int_{|u|>1} H(du) < \infty.$$

In particular, the cumulant of the process with finite variation has the form:

$$\Gamma(\lambda) = i\lambda a + \int_{-\infty}^{+\infty} (e^{i\lambda u} - 1) H(du)$$

with the spectral measure satisfying condition

$$\int_{|u|\le 1} |u| H(du) < \infty.$$

The finite spectral measure on the real line: $H(R) = h < \infty$ determines the *compound Poisson process*, which can be determined as follows

$$\xi(t) = \sum_{k=1}^{\nu(t)} \zeta_k,$$

where ζ_k, $k \geq 1$, are mutually independent identically distributed random variables with the distribution function

$$\Phi(B) = H(B)/h = \mathcal{P}\{\zeta_k \in B\}.$$

The process $\nu(t)$, $t \geq 0$ is the pure Poisson process with the rate parameter h:

$$\mathcal{P}\{\nu(t) = n\} = e^{-ht}(ht)^n/n!, \quad n \geq 0.$$

The cumulant of the compound Poisson process has the form

$$\Gamma(\lambda) = h \int_{-\infty}^{\infty} (e^{i\lambda u} - 1)\Phi(du). \tag{4.3}$$

The process with stationary independent increments possesses Markov property [19], that is, the transitional distributions are generated by the generating operator of the Markov semi-group:

$$\Gamma_t \varphi(u) := E\varphi(u + \xi(t)), \quad t \geq 0. \tag{4.4}$$

Lemma 4.1. [19] The generator Γ of the semi-group (4.4) has the following representation

$$\Gamma\varphi(u) = \int_{-\infty}^{+\infty} e^{i\lambda u}\Gamma(\lambda)\tilde{\varphi}(\lambda)d\lambda \tag{4.5}$$

for $\varphi(u) = \int_{-\infty}^{+\infty} e^{i\lambda u}\tilde{\varphi}(\lambda)d\lambda$, where $\tilde{\varphi}(\lambda)$ and $\lambda^2\tilde{\varphi}(\lambda)$ are integrable functions.

Proof of Lemma 4.1. Let us consider semi-group (4.4) using the definition (4.1)

$$\begin{aligned} \Gamma_t\varphi(u) &= E\varphi(u + \xi(t)) \\ &= E\int_{-\infty}^{+\infty} e^{i\lambda(u+\xi(t))}\tilde{\varphi}(\lambda)d\lambda \\ &= \int_{-\infty}^{+\infty} e^{i\lambda u + t\Gamma(\lambda)}\tilde{\varphi}(\lambda)d\lambda. \end{aligned}$$

Note, that $\Gamma(\lambda) = O(\lambda^2)$ as $|\lambda| \to \infty$ (see (4.2)). Hence, the last integral is convergent uniformly on t. So, we can get the derivative

$$d\Gamma_t\varphi(u)/dt = \int_{-\infty}^{+\infty} e^{i\lambda u + t\Gamma(\lambda)}\Gamma(\lambda)\tilde{\varphi}(\lambda)d\lambda.$$

By the evolutional equation for the semi-group there is

$$d\Gamma_t\varphi(u)/dt = \Gamma_t\Gamma\varphi(u).$$

Comparing the latest two formulas we get (4.5).

The meaning of representation (4.5) is that the cumulant $\Gamma(\lambda)$ is the *symbol* of the generator Γ.

In particular case of the *shift process* $\xi(t) = at$, the corresponding generator $\Gamma\varphi(u) = a\varphi'(u)$ because

$$\int_{-\infty}^{+\infty} e^{i\lambda u} i\lambda a \tilde{\varphi}(\lambda)d\lambda = a\varphi'(u).$$

It is well-known that the standard (Wiener) process with the cumulant

$$\Gamma(\lambda) = -\sigma^2\lambda^2/2$$

has the generator

$$\Gamma\varphi(u) = \sigma^2\varphi''(u)/2.$$

Corollary 4.1. The generator of the semi-group (4.4) with the cumulant (4.2) has the following representation

$$\begin{aligned}\Gamma\varphi(u) &= a\varphi'(u) - \sigma^2\varphi''(u)/2 \\ &+ \int_{-\infty}^{+\infty} [\varphi(u+v) - \varphi(u) - v\varphi'(u)I_{(|u|\leq 1)}]H(dv).\end{aligned}$$

In particular case of compound Poisson process with the cumulant (4.3) the generator has the form

$$\Gamma\varphi(u) = h\int_{-\infty}^{+\infty} [\varphi(u+v) - \varphi(u)]\Phi(dv).$$

Stochastic additive functionals. Let $\kappa(t)$, $t \geq 0$, be a Markov jump process on a standard phase space $(X, \mathcal{X})$ with a locally finite number of jumps [18]. Introduce the moment of the first jump

$$\tau = \min\{t : \kappa(t) \neq \kappa(0)\}.$$

Definition 4.1. *The stochastic additive functional* $\zeta(t), t \geq 0$, which is taking real values, is defined by the following property: the conditional distribution of the two component process $\zeta(\tau + t) - \zeta(\tau)$, $\kappa(\tau + t)$; $t \geq 0$, for a fixed $\kappa(\tau)$ is independent from and identically distributed with the process $\zeta(t)$, $\kappa(t)$; $0 \leq t \leq \tau$, under condition $\zeta(0) = 0$, $\kappa(0) = \kappa(\tau)$.

The stochastic additive functional with right-continuous trajectories has the following properties [18]:

(a) The breaking process $\zeta(t), 0 \leq t \leq \tau$, is a process with stationary independent increments;

(b) The random variables $\zeta(\tau) - \zeta(\tau - 0)$, $\kappa(\tau)$, are independent from $\zeta(t)$, $0 \leq t \leq \tau$, for a fixed $\kappa(0)$.

Analytically the stochastic additive functional is determined by the following parameters: the *cumulant* $\Gamma(\lambda; x)$, $x \in X$, of the process $\xi(t; x)$, $x \in X$, $t \geq 0$, with stationary independent increments:

$$E \exp[i\lambda\xi(t; x)] = \exp[t\Gamma(\lambda; x)];$$

the *intensity of the first jump* τ:

$$\mathcal{P}\{\tau > t/\kappa(0) = x\} = \exp[-q(x)t];$$

the *distribution function of jumps*:

$$\mathcal{P}\{\eta(x, y) < u\} = G_{xy}(u), \quad x, y \in X, \quad u \in R.$$

The breaking process $\zeta(t)$, $0 \leq t \leq \tau$, is described as follows

$$E_x[\exp[i\lambda\zeta(t)]I(\tau > t)] = \exp[t(\Gamma(\lambda; x) - q(x))],$$

where E_x means, as usual, conditional expectation under $\kappa(0) = x$.

The jump $\zeta(\tau) - \zeta(\tau-)$, is described as follows

$$\mathcal{P}\{\zeta(\tau) - \zeta(\tau-) < u, \kappa(\tau) \in B/\kappa(0) = x\} = \int_B P(x, dy)G_{xy}(u),$$

where the stochastic kernel $P(x, dy)$ determines the transition probabilities of the imbedded Markov chain $\kappa_n := \kappa(\tau_n), n \geq 0$:

$$P(x, B) = \mathcal{P}\{\kappa_{n+1} \in B/\kappa_n = x\}.$$

So, the stochastic additive functional $\zeta(t), t \geq 0$, of the Markov jump process $\kappa(t)$, $t \geq 0$, is determined by the intensity parameter $q(x)$, $x \in X$, the stochastic kernel $P(x, B)$, $x \in X$, $B \in \mathcal{X}$, the distribution functions G_{xy}, $x, y \in X$, and the cumulants $\Gamma(\lambda; x), x \in X$.

The following stochastic representation

$$\zeta(t) = \int_0^t \xi(ds, \kappa(s)) + \sum_{k=1}^{\nu(t)} \eta(\kappa_{k-1}, \kappa_k),$$

where $\nu(t) := \max\{n : \tau_n \leq t\}$ is the counting process of the Markov jump process $\kappa(t), t \geq 0$, arises from the properties of the stochastic additive functional mentioned above.

The process $\xi(t; x)$ and the random variables $\eta(x, y)$ are mutually independent for every finite set of $x, y \in X$.

Generator of stochastic additive functional. The two component random process $\zeta(t), \kappa(t); t \geq 0$, is a Markov one and can be defined by the generator on the Banach space $\mathcal{B}_{R \times X}$ of bounded real-valued functions $\varphi(u, x)$.

Lemma 4.2. The generator L of the Markov process $\zeta(t), \kappa(t); t \geq 0$, is determined by the form

$$L\varphi(u, x) = q(x) \int_X P(x, dy)$$

$$\times \int_{-\infty}^{+\infty} G_{xy}(dv)[\varphi(u + v, y) - \varphi(u, x)] + \Gamma(x)\varphi(s, x), \qquad (4.6)$$

where $\Gamma(x)$, $x \in X$, is the collection of generators of processes with stationary independent increments $\xi(t; x), x \in X$, with the cumulants $\Gamma(\lambda; x), x \in X$. The function $\varphi(u, x)$ is considered on the common domain $\mathcal{B}_0 := \cap_{x \in X} \mathcal{D}_{\Gamma(x)}$.

Proof of Lemma 4.2. First we calculate the asymptotic representation of the conditional expectation as $t \to 0$.

$$\begin{aligned} E_{u,x}\varphi(\zeta(t), \kappa(t)) &= E_{u,x}[\varphi(\zeta(t), x)I(\tau > t)] \\ &\quad + E_{u,x}[\varphi(\zeta(t), \kappa(t))I(\tau \leq t)]. \end{aligned} \qquad (4.1.7)$$

The first term has the following asymptotic representation

$$E_{u,x}[\varphi(\zeta(t), x)I(\tau > t)]$$

$$= \varphi(u, x)(1 - q(x)t) + E_{u,x}[\varphi(\zeta(t), x) - \varphi(u, x)] + o(t). \qquad (4.1.8)$$

The second term in (4.7) we transform as follows

$$E_{u,x}[\varphi(\zeta(t), \kappa(t))I(\tau \le t)] = E_{u,x}[\varphi(\zeta(\tau) - \zeta(\tau-), \kappa(\tau))I(\tau \le t)]$$

$$+E_{u,x}[(\varphi(\zeta(t), \kappa(t)) - \varphi(\zeta(\tau), \kappa(\tau)))I(\tau \le t)]$$

$$+E_{u,x}[(\varphi(\zeta(\tau), \kappa(\tau)) - \varphi(\zeta(\tau) - \zeta(\tau-), \kappa(\tau)))I(\tau \le t)].$$

It can be verified that the second and third term of this formula are $o(t)$. Therefore the second term in (4.7) is represented as follows

$$E_{u,x}[\varphi(\zeta(t), \kappa(t))I(\tau \le t)] = E_{u,x}[\varphi(\zeta(\tau) - \zeta(\tau-), \kappa(\tau))I(\tau \le t)] + o(t)$$

$$= tq(x) \int_X P(x, dy) \int_{-\infty}^{+\infty} G_{xy}(dv)\varphi(u + v, y) + o(t). \tag{4.9}$$

Combining (4.7)–(4.9) we get (4.6).

Let us extract the generator Q of the Markov jump process $\kappa(t)$, $t \ge 0$, of the generator L.

Corollary 4.2. The generator of the Markov process $\zeta(t), \kappa(t); t \ge 0$, has the following representation:

$$L = Q + \Gamma(x) + Q_0[G_x - I], \tag{4.10}$$

where

$$Q\varphi(u, x) := q(x) \int_X P(x, dy)[\varphi(u, y) - \varphi(u, x)],$$

$$Q_0[G_x - I]\varphi(u, x) := q(x) \int_X P(x, dy) \int_{-\infty}^{+\infty} G_{xy}(dv)[\varphi(u + v, y) - \varphi(u, y)].$$

Corollary 4.3. The continuous stochastic additive functional

$$\zeta^c(t) := \int_0^t \xi(ds, \kappa(s))$$

is determined by the generator

$$L = Q + \Gamma(x).$$

Corollary 4.4. The jump stochastic additive functional

$$\zeta^j(t) = \sum_{k=1}^{\nu(t)} \eta(\kappa_{k-1}, \kappa_k)$$

is determined by the generator

$$L = Q + Q_0[G_x - I].$$

Corollary 4.5. The stochastic integral functional

$$\zeta^I(t) := \int_0^t a(\kappa(s))ds$$

is determined by the generator

$$L = Q + \Gamma(x), \quad \Gamma(x)\varphi(u) := a(x)d\varphi(u)/du.$$

4.2 Storage Processes

The various types of stochastic systems can be described as *the switched dynamic systems*.

Let $C(u, x)$, $u \in R$, $x \in X$ be a function continuously differentiable with respect to u with bounded first derivative $C_u' := \partial C(u, x)/\partial u$.

It is well-known that, for such a function, there exists a unique solution of the dynamic equation

$$dU^x(t)/dt = C(U^x(t), x),$$

$$U^x(0) = U_0 \tag{4.11}$$

for every fixed value $x \in X$.

Definition 4.2. *The storage process* $U(t)$, $t \geq 0$ is defined as a solution of the integral equation

$$U(t) = U_0 + \int_0^t C(U(s), \kappa(s)) ds + \sum_{k=1}^{\nu(t)} a(\kappa(t)), \qquad (4.12)$$

where $a(x)$, $x \in X$ is a bounded real-valued function on X.

For simplicity, we consider real-valued storage processes only. Without any loss of generality the finite-dimensional Euclidean space $\mathbb{R}^N$ can be taken as a space of values for storage processes.

The correctness of the storage process definition is a consequence of the regularity of the Markov jump process $\kappa(t)$ and the unique solvability of dynamic equation (4.11) for every fixed $x \in X$. We may define the storage process $U(t)$ as a solution of the Cauchy problem

$$dU(t)/dt = C(U(t), \kappa_k),$$

$$U(\tau_k) = U(\tau_k-) + a(\kappa_k)$$

on every interval $[\tau_k, \tau_{k+1})$, $k \geq 0$, $\tau_0 = 0$.

Denote by

$$U^x(t) := U^x(t; U_0)$$

the solution of the Cauchy problem (4.11). Then the storage process can be defined recurrently

$$U(t) = U^{\kappa_k}(t; U(\tau_k)),$$

$$\tau_k \leq t < \tau_{k+1}, \quad k \geq 0, \qquad (4.13)$$

and

$$U(\tau_{k+1}) = U^{\kappa_k}(\tau_{k+1}-; U(\tau_k)) + a(\kappa_k). \qquad (4.14)$$

By using a point process $\tau(t) := \tau_{\nu(t)}$, the storage process can be represented, instead of (4.13)–(4.14), by the solutions of the Cauchy problem (4.11) as

$$U(t) = U^{\kappa(t)}(t; U(\tau(t))), \quad t \geq 0.$$

Generator of storage process. The two-component random process $U(t)$, $\kappa(t)$, $t \geq 0$, where $U(t)$ is the storage process determined by the integral equation (4.12), is a Markov process.

Lemma 4.3. The generator L of the Markov process $U(t), \kappa(t), t \geq 0$, is determined by the relation

$$L\varphi(u, x) = Q\varphi(u, x) + C(x)\varphi(u, x) + Q_0[A(x) - I]\varphi(u, x), \qquad (4.15)$$

where Q is a generator of the switching Markov process $\kappa(t)$, and $C(x)$, $x \in X$ are generators of the dynamic deterministic system (4.11):

$$C(x)\varphi(u, x) := C(u, x)\varphi_u'(u, x). \qquad (4.16)$$

The operators $A(x)$, $x \in X$ act as shift operators:

$$A(x)\varphi(u, x) := \varphi(u + a(x), x). \qquad (4.17)$$

Note that the operator Q_0 was defined in Corollary 4.2.

The operator L is defined on the function $\varphi(u, x)$ continuously differentiable with respect to u with bounded first derivative.

Proof of Lemma 4.3. Consider, for a small $t > 0$, the conditional expectation

$$E_{u,x}[\varphi(U(t), \kappa(t)) - \varphi(u, x)] = E_{u,x}[\varphi(u, \kappa(t)) - \varphi(u, x)]$$

$$+ E_{u,x}[(\varphi(U(t), x) - \varphi(u, x))I(\tau > t)]$$

$$+ E_{u,x}[(\varphi(U(t), \kappa(t)) - \varphi(u, \kappa(t)))I(\tau \leq t)]. \qquad (4.18)$$

The first term in (4.18) gives us the generator Q of the Markov process $\kappa(t)$. The second term in (4.18) has the following representation

$$E_{u,x}[(\varphi(U(t), x) - \varphi(u, x))I(\tau > t)]$$

$$= E_{u,x}[\varphi(u + \int_0^t C(U(v), x)dv, x) - \varphi(u, x)] + o(t)$$

$$= tC(u, x)\varphi_u'(u, x) + o(t) = tC(x)\varphi(u, x) + o(t).$$

The third term in (4.18) can be transformed as follows

$$E_{u,x}[(\varphi(U(t), \kappa(t)) - \varphi(u, \kappa(t)))I(\tau \le t)]$$
$$= E_{u,x}[(\varphi(u + a(x), \kappa(t)) - \varphi(u, \kappa(t)))I(\tau \le t)] + o(t)$$
$$= tQ_0[A(x) - I]\varphi(u, x) + o(t).$$

If we gather the asymptotic representations of terms in (4.18), we obtain (4.15)–(4.17).

Corollary 4.6. A continuous storage process defined by the solu tion of the evolutional equation

$$dU(t)/dt = C(U(t), \kappa(t)), \ U(0) = U_0,$$

can be determined by the generator

$$L = Q + C(x)$$

of the two-component Markov process $U(t), \kappa(t), \ t \ge 0$.

Corollary 4.7. An impulse storage process defined by the relati on

$$U(t) = \sum_{k=1}^{\nu(t)} a(\kappa_k)$$

can be determined by the generator

$$L = Q + Q_0[A(x) - I]$$

of the two–component Markov process $U(t), \kappa(t), \ t \ge 0$.

Comparing the generators of the stochastic additive functional (4.10) and storage process (4.15) we can conclude that there exists a certain general scheme for an evolutional stochastic system, which contains a stochastic additive functional and storage process as particular cases. Indeed, such an abstract scheme is *a random evolution.*

4.3   Random evolution

Let $\kappa(t)$, $t \geq 0$ be a switching Markov process on a standard phase space $(X, \mathcal{X})$. Let a family of generators $\Gamma(x)$, $x \in X$ of strictly continuous contractive semi-groups $\Gamma_t(x)$, $t \geq 0$, $x \in X$ and a set of bounded linear operators $D(x)$, $x \in X$ on a Banach space $\mathcal{B}$ be given.

Definition 4.3. A random evolution is defined as an operator va lued random process $V(t)$, $t \geq 0$ on the Banach space $\mathcal{B}$ that satisfy the integral equation

$$V(t) = I + \int_0^t \Gamma(\kappa(s))V(s)ds + \sum_{k=1}^{\nu(t)}[D(\kappa_k) - I]V(\tau_k-). \qquad (4.19)$$

Using the Markov renewal process $(\kappa_n, \theta_n; \, n \geq 0)$, which induces the Markov process $\kappa(t) = \kappa_{\nu(t)}$, $\nu(t) := \max\{n : \tau_n \leq t\}$, $\tau_n := \sum_{k=1}^n \theta_k$, $n \geq 0$, $\tau_0 = 0$, and the auxiliary defect processes $\gamma(t) := t - \tau(t)$, $\tau(t) := \tau_{\nu(t)}$, the random evolution $V(t)$ can be represented in the form

$$V(t) = \Gamma_{\gamma(t)}(\kappa(t)) \prod_{k=1}^{\nu(t)} D(\kappa_k)\Gamma_{\theta_k}(\kappa_{k-1}). \qquad (4.20)$$

The semi-groups $\Gamma_t(x)$, $t \geq 0$ determine a continuous part of the evolution whereas the linear operators $D(x)$ define an impulse part of the evolution. The continuous evolution is realized by (4.20) under the condition $D(x) \equiv I$. By using (4.19) it is easy to verify that a continuous evolution satisfies the integral equation

$$V^c(t) = I + \int_0^t \Gamma(\kappa(s))V^c(s)ds,$$

which is obviously equivalent to the evolutional equation

$$dV^c(t)/dt = \Gamma(\kappa(t))V^c(t), \quad V^c(0) = I.$$

Note that the equivalence of the integral and the evolutional equation for a random evolution is a consequence of the integral equation for the semi-group [5]

$$\Gamma_t(x) = I + \int_0^t \Gamma(x)\Gamma_s(x)ds. \qquad (4.21)$$

Considering the random evolution (4.19) on the intervals $[\tau_k, \tau_{k+1})$ we get the product representation (4.20). The first multiplier in (4.20) is a solution of the residual equation

$$V(t) = V(\tau(t)) + \int_{\tau(t)}^{t} \Gamma(\kappa(s))V(s)ds.$$

Let $\varphi(x) : X \to \mathcal{B}$ be a function of $x \in X$ taking values in a Banach space $\mathcal{B}$ where the random evolution $V(t)$, $t \geq 0$ acts.

Lemma 4.4. The mean value of the random evolution

$$
\begin{aligned}
U(t, x) &:= E_x[V(t)\varphi(\kappa(t))] \\
&:= E[V(t)\varphi(\kappa(t))/\kappa(0) = x]
\end{aligned}
$$

is determined by a solution of *the Markov renewal equation*

$$U(t, x) - q(x) \int_0^t e^{-q(x)(t-s)} ds \int_X P(x, dy)D(y)\Gamma_{t-s}(x)U(s, y)$$

$$= e^{-q(x)t}\Gamma_t(x)\varphi(x) \qquad (4.22)$$

or, in the equivalent form, by a solution of the evolutional equation

$$dU(t, x)/dt = [Q + \Gamma(x) + Q_0[D(x) - I]]U(t, x),$$

$$U(0, x) = \varphi(x). \qquad (4.23)$$

Proof of Lemma 4.4. By using the first moment of a jump

$$\tau := \inf\{t : \kappa(t) \neq \kappa(0)\}$$

we can write

$$U(t, x) = E_x[V(t)\varphi(\kappa(t))I(\tau > t)] + E_x[V(t)\varphi(\kappa(t))I(\tau \leq t)]. \qquad (4.24)$$

The first term in (4.24) has the obvious representation

$$E_x[V(t)\varphi(\kappa(t))I(\tau > t)] = e^{-q(x)t}\Gamma_t(x)\varphi(x). \qquad (4.25)$$

Using the strong Markov property of the process $\kappa(t)$ and taking into account the product representation (4.20) for a random evolution we get the following representation for the second term in (4.24)

$$E_x[V(t)\varphi(\kappa(t))I(\tau \leq t)] =$$
$$q(x)\int_0^t e^{-q(x)s}ds \int_X P(x,dy)D(y)\Gamma_s(x)U(t-s,y).$$

If we change the variable $s = t - s'$, we obtain

$$E_x[V(t)\varphi(\kappa(t))I(\tau \leq t)]$$

$$= q(x)\int_0^t e^{-q(x)(t-s)}ds \int_X P(x,dy)D(y)\Gamma_{t-s}(x)U(s,y). \tag{4.26}$$

The equalities (4.24)–(4.26) give us the Markov renewal equation (4.22). If we differentiate the equation (4.22) with respect to t and use the evolution equation for the semi-group

$$d\Gamma_t(x)/dt = \Gamma(x)\Gamma_t(x),$$

after an ordinary transformation we obtain the evolution equation (4.23).

Remark 4.1. The $\mathcal{B}$-valued process $V(t)\varphi(\kappa(t))$, $t \geq 0$ is determined by the generator

$$L\varphi(x) = [Q + \Gamma(x) + Q_0[D(x) - I]]\varphi(x)$$

and has the martingale characterization of the following form

$$V(t)\varphi(\kappa(t)) - \int_0^t LV(s)\varphi(\kappa(s))ds = \mu_t,$$

where μ_t, $t \geq 0$ is a martingale adapted to the filtration

$$F_t = \sigma\{\kappa(s),\, 0 \leq s \leq t\}.$$

Interpretation of random evolution. The stochastic systems menti oned in Sections 4.1 and 4.2 have a certain interpretation as random evolutions with definite local characteristics.

Lemma 4.5. The storage process

$$U(t) = U + \int_0^t C(U(s), \kappa(s))ds + \sum_{k=1}^{\nu(t)} a(\kappa_k) \tag{4.27}$$

can be interpreted as a random evolution on a Banach space $\mathcal{B}_R$ of bounded real valued functions $\varphi(u)$, $u \in R$ with sup-norm

$$\|\varphi\| := \sup_{u \in R} |\varphi(u)|$$

in the following manner

$$V(t)\varphi(u) := \varphi(u + U(t)) \tag{4.28}$$

with semi-groups

$$\Gamma_t(x)\varphi(u) = \varphi(u + \int_0^t C(U(s), x)ds)$$

induced by generators

$$\Gamma(x)\varphi(u) = C(u, x)\varphi'(u) \tag{4.29}$$

and with the set of bounded linear operators

$$D(x)\varphi(u) = \varphi(u + a(x)). \tag{4.30}$$

Proof of Lemma 4.5. If we consider the random evolution (4.27) on the intervals $[\tau_k, \tau_{k+1})$, $k \geq 0$, then, proceeding by induction, we get

$$\begin{aligned} V(t)\varphi(u) &= \varphi(u + \int_0^t C(U(s), \kappa_0)ds) \\ &= \Gamma_t(\kappa_0)\varphi(u), \quad 0 \leq t < \tau_1, \end{aligned}$$

and

$$\begin{aligned} V(t)\varphi(u) &= V(\tau_k)\varphi(u + \int_{\tau_k}^t C(U(s), \kappa_k)ds) \\ &= V(\tau_k)\Gamma_{t-\tau_k}(\kappa_k)\varphi(u), \quad \tau_k \leq t < \tau_{k+1}. \end{aligned}$$

In addition, by definition (4.28) and (4.30)

$$\begin{aligned} V(\tau_k)\varphi(u) &= V(\tau_k-)\varphi(u + a(\kappa_k)) \\ &= V(\tau_k-)D(\kappa_k)\varphi(u). \end{aligned}$$

Moreover, on the residual interval $[\tau(t), t)$ we have

$$
\begin{aligned}
V(t)\varphi(u) &= V(\tau(t))\varphi(u + \int_{\tau(t)}^{t} C(U(s),\kappa(t))ds) \\
&= \Gamma_{\gamma(t)}(\kappa(t))\varphi(u)V(\tau(t)).
\end{aligned}
$$

Gathering all these representations, we get the product form (4.20) of a random evolution with local characteristics (4.29) and (4.30).

Lemma 4.6. The stochastic additive functional

$$
\zeta(t) = \int_{0}^{t} \xi(ds, \kappa(s)) + \sum_{k=1}^{\nu(t)} \eta(\kappa_k) \tag{4.31}
$$

can be interpreted as a random evolution on the Banach space in the following manner

$$
V(t)\varphi(u) = E[\varphi(u + \zeta(t))/\kappa(s), 0 \le s \le t]
$$

with semi-groups

$$
\Gamma_t(x)\varphi(u) = E\varphi(u + \xi(t; x))
$$

induced by generators $\Gamma(x)$, $x \in X$ of processes with stationary independent increments $\xi(t; x)$, $x \in X$ with cumulants $\Gamma(\lambda; x)$, $x \in X$.

The impulse part of evolution (4.31) is determined by bounded linear operators

$$
D(x)\varphi(u) = E\varphi(u + \eta(x)).
$$

Proof of Lemma 4.6 differs from that of Lemma 4.4 only in that t he random evolution (4.31) on intervals $[\tau_k, \tau_{k+1})$, $k \ge 0$, is stochastic while the random evolution (4.27) is deterministic.

Remark 4.2. The abstract form (4.19) of a random evolution takes

different form than in (4.27) and (4.31) only after one has given concrete expressions for the local characteristics $\Gamma(x)$ and $D(x)$, $x \in X$.

4.4 Ergodic average and diffusion approximation of random evolutions

The general idea of simplification of a stochastic system described in Section 4.1 and 4.2 is considered here for random evolutions in a series scheme with a small series parameter $\varepsilon > 0$.

We exhibit the ergodic average algorithms for stochastic evolutions on increasing time intervals in an ergodic random Markov medium.

A random evolution in the series scheme with a small series parameter $\varepsilon > 0$ is defined as the solution of the operator equation in the Banach space $\mathcal{B}$ (cf. (4.19))

$$V^\varepsilon(t) = I + \int_0^t \Gamma(\kappa(s/\varepsilon))V^\varepsilon(s)ds + \sum_{n=1}^{\nu(t/\varepsilon)} [D^\varepsilon(\kappa_n^\varepsilon) - I]V^\varepsilon(\tau_n^\varepsilon-). \qquad (4.32)$$

As was shown in Section 4.3 random evolutions can always be identified with some stochastic evolution described by the two component Markov process. Therefore, without loss of generality, a random evolution (4.32) in the series scheme can be considered in the following form.

$$V^\varepsilon(t)\varphi(u) = E[\varphi(u + \zeta^\varepsilon(t))/\kappa(s), \ 0 \le s \le t]$$

with some stochastic evolutional process $\zeta^\varepsilon(t)$ ($\zeta^\varepsilon(0) = 0$).

The weak convergence of random evolutions

$$V^\varepsilon(t) \Rightarrow \widehat{V}(t) \quad \text{as} \quad \varepsilon \to 0$$

takes place on the dense domain $\mathcal{B}_0$ in the Banach space $\mathcal{B}$ of functions $\varphi(u)$ or, in the phase space of trajectories of stochastic evolutional processes the weak convergence

$$\zeta^\varepsilon(t) \Rightarrow \zeta(t) \quad \text{as} \quad \varepsilon \to 0$$

takes place. It is supposed that the domain of weak convergence $\mathcal{B}_0$ contains the set of separating and convergence determining functions [4].

Theorem 4.1. (*Ergodic average*) Let the following conditions be satisfied:

C1: The intensity of jumps of a Markov process $\kappa(t)$, $t \ge 0$ is bounded and strictly positive

$$0 < q_0 \le q(x) \le q_1 < +\infty;$$

C2: The imbedded Markov chain κ_n, $n \geq 0$ with a stochastic kernel $P(x, B)$ is uniformly ergodic with a stationary distribution $\rho(B)$, $B \in \mathcal{X}$;

C3: There is the asymptotic expansion

$$D^\varepsilon(x)\varphi = [I + \varepsilon D(x) + \varepsilon\theta_\varepsilon(x)]\varphi \qquad (4.33)$$

for all $\varphi \in \mathcal{B}_0$ of a common domain of operators $D(x)$ and $\theta_\varepsilon(x)$, $x \in X$, dense in $\mathcal{B}$. The operators $\theta_\varepsilon(x)$ satisfy the condition

$$\|\theta_\varepsilon(x)\varphi\| \to 0 \quad \text{as } \varepsilon \to 0, \quad \varphi \in \mathcal{B}_0;$$

C4: The generators $\Gamma(x)$ and $D(x)$, $x \in X$ and the averaged operators

$$\Gamma = \int_X \pi(dx)\Gamma(x), \ D = q \int_X \rho(dx)D(x) \qquad (4.34)$$

as well as the operators

$$[\Gamma(x) + q(x)D(x)]R_0[\pi(x) + q(x)D(x)], \ x \in X$$

are closed and have common domain $\mathcal{B}_0$.

Then the weak convergence

$$V^\varepsilon(t) \Rightarrow \widehat{V}(t) \quad \text{as } \varepsilon \to 0$$

takes place. The averaged evolution $\widehat{V}(t)$ is determined by a solution of the following equation

$$\widehat{V}(t) = I + \int_0^t [\Gamma + D]\widehat{V}(s)ds$$

or, in another form,

$$d\widehat{V}(t)/dt = [\Gamma + D]\widehat{V}(t), \ \widehat{V}(0) = I.$$

The proof of this theorem is given in Section 4.8.

Note that the averaged evolution is continuous and defined by the generator Γ averaged over the stationary distribution $\pi(dx)$ of the Markov process $\kappa(t)$, $t \geq 0$, and by the operator D averaged over the stationary distribution $\rho(dx)$ of the embedded Markov chain κ_n, $n \geq 0$, with the scale factor

$q = \int_X \pi(dx)q(x)$. The heuristic interpretation of the second formula in (4.34) is as follows. According to the definition of a random evolution (4.32) the small increments of jump components are occurred in renewal moments τ_n, $n \geq 1$. Therefore, the increments of jumps are averaged over the stationary distribution of the embedded Markov chain defined by the frequency of states. But the final averaged continuous rate D acts in continuous time. The transfer from the jump evolution to a continuous one is necessary to account the averaged rate of jump intensity q as the scale factor.

The *ergodic average algorithm for a stochastic system,* which can be described by the random evolution, is based on Theorem 4.1.

Proposition 4.1. The ergodic average algorithm is described by the following formulas:
the averaged generator of a continuous component

$$\Gamma = \int_X \pi(dx)\Gamma(x);$$

the asymptotic expansion of a jump component

$$D_T(x) = I + D(x)/T + \theta_\varepsilon(x)/T \quad \text{as } T \to \infty$$

with a time scale parameter T and the averaged operator of the jump component

$$D = q \int_X \rho(dx)D(x),$$

$$q = \int_X \pi(dx)q(x).$$

The small series parameter $\varepsilon = 1/T$. For a given random evolution $V(t)$, $t \geq 0$, the ergodic average algorithm establishes the approximate equality

$$V(Tt) \simeq \hat{V}(t).$$

The problem arises how to choose the scale parameter T in order to get the error in this approximation small enough.

Ergodic diffusion approximation. The averaging effect can be trivial if the averaged operators Γ and D vanish: $\Gamma = D = 0$.

Under such a balance condition (and without one) the opportunity arises to study fluctuations of a random evolution relative to the averaged evolution. It is natural to expect that fluctuations can be described by diffusion processes.

Note that, in order to get a nontrivial averaged effect under a certain balance condition, we have to consider a random evolution in the series scheme with a small series parameter $\varepsilon > 0$ of the following form (cf. (4.32))

$$V^\varepsilon(t) = I + \int_0^t \Gamma^\varepsilon(\kappa(s/\varepsilon^2))V^\varepsilon(s)ds + \sum_{n=1}^{\nu(t/\varepsilon^2)} [D^\varepsilon(\kappa_n^\varepsilon) - I]V^\varepsilon(\varepsilon\tau_n^\varepsilon-)$$

Theorem 4.2 (*Ergodic Diffusion Approximation*) Let the conditions C1 and C2 of Theorem 4.1 be satisfied along with the following additional conditions:

C1: There are asymptotic expansions in $\mathcal{B}_0$

$$D^\varepsilon(x)\varphi = [I + \varepsilon D(x) + \varepsilon^2 D_0(x) + \varepsilon^2 \theta_{\varepsilon 1}(x)]\varphi \qquad (4.35)$$

and

$$\Gamma^\varepsilon(x)\varphi = [\Gamma(x) + \varepsilon\Gamma_0(x) + \varepsilon\theta_{\varepsilon 2}(x)]\varphi, \qquad (4.36)$$

where operators $\theta_{\varepsilon k}(x)$ satisfy the negligible condition

$$\|\theta_{\varepsilon k}(x)\varphi\| \to 0 \quad \text{as } \varepsilon \to 0, \quad k = 1, 2;$$

C2: The balance condition is true on $\mathcal{B}_0$

$$\int_X \pi(dx)[\Gamma(x) + q(x)D(x)]\varphi = 0. \qquad (4.37)$$

C3: The operators

$$L(x) = L_0(x) + L_1(x),$$
$$L_0(x) = \Gamma_0(x) + q(x)D_0(x),$$
$$L_1(x) = [\Gamma(x) + q(x)D(x)]R_0[\Gamma(x) + q(x)D(x)] \qquad (4.38)$$

are closed in the common domain $\mathcal{B}_0$; and the averaged operator

$$L = \int_X \pi(dx)L(x) \qquad (4.39)$$

is a generator of the strictly continuous semi-group.

Then the weak convergence

$$V^\varepsilon(t) \Rightarrow \widehat{W}(t) \text{ as } \varepsilon \to 0$$

takes place. The averaged evolution $\widehat{W}(t)$ is defined by a solution of the martingale problem

$$[\widehat{W}(t) - I - \int_0^t L\widehat{W}(s)ds]\varphi = \mu_t, \ \varphi \in \mathcal{B}_0$$

for the generator L and the martingale μ_t, $t \geq 0$ with respect to the filtration

$$F_t = \sigma\{\widehat{W}(s)\varphi; \ 0 \leq s \leq t\}, \ t \geq 0.$$

The proof of Theorem 4.2 is given in Section 4.8.

Remark 4.3. The limiting evolution $\widehat{W}(t)$ is not exac tly a diffusion process just as the operator L can be considered as a generator of a diffusion process, which always takes place in applications (see Section 4.6).

4.5 Splitting average and diffusion approximation of random evolution

The average effect essentially depends on the ergodic property of a random medium. Using the splitting merging algorithm given in Section 3.5 we consider the splitting averaging and diffusion approximation of random evolutions.

The switching Markov process $\kappa^\varepsilon(t), t \geq 0$, dependent on the series parameter $\varepsilon > 0$ is considered in the following form. The generator of process $\kappa^\varepsilon(t), t \geq 0$, is represented as follows

$$Q^\varepsilon = Q + \varepsilon Q_1,$$

where Q is the generator of the support Markov process $\kappa^0(t)$, $t \geq 0$, which is uniformly ergodic in the splitting phase space

$$X = \cup_{k=1}^N X_k, \quad X_k \cup X_{k'} = \emptyset, \ k \neq k',$$

with the stationary distributions $\pi_k(dx)$, $\pi_k(X_k) = 1$, $1 \leq k \leq N$. So that the operator Q on the Banach space $\mathcal{B}_X$ is reducible-invertible (Section 3.1) with zero-space $N_Q = \{\varphi : \ Q\varphi = 0\}$ of the dimensionality $N > 1$. The projector Π on the zero-space N_Q is determined by the relation

$$\Pi\varphi(x) = \sum_{k=1}^{N} \hat{\varphi}_k 1_k(x),$$

$$\hat{\varphi}_k := \int_{X_k} \pi_k(dx)\varphi(x).$$

Here, as usual,

$$1_k(x) = \begin{cases} 1, & x \in X_k, \\ 0, & x \notin X_k. \end{cases}$$

The main assumption is that the merged operator $\hat{Q}$, given by the relation (Section 3.2.)

$$\Pi Q_1 \Pi = \hat{Q}\Pi,$$

determines the generator matrix

$$\hat{Q} = [q_{kr}; 1 \leq k, r \leq N]$$

of the merged Markov process $\hat{\kappa}(t)$, $t \geq 0$, on the merged phase space

$$\widehat{X} = \{1, 2, \ldots, N\}.$$

Note that due to Theorem 3.6 the weak convergence

$$k(\kappa^\varepsilon(t/\varepsilon)) \Rightarrow \hat{\kappa}(t) \quad \text{as } \varepsilon \to 0, \tag{4.40}$$

takes place.

Theorem 4.3. (*Splitting average*) Let the conditions C1 and C3 of Theorem 4.1 be valid and, in addition, the following conditions be fulfilled:
C1: The weak convergence (4.40) takes place.
C2: The averaged operators

$$\widehat{\Gamma}(k) = \int_{X_k} \pi_k(dx)\Gamma(x)$$

and

$$\widehat{D}(k) = q_k \int_{X_k} \rho_k(dx)D(x),$$

$$q_k = \int_{X_k} \pi_k(dx)q(x)$$

are closed with dense domain $\mathcal{B}_0 \subset \mathcal{B}$.

Then the random evolution $V^\varepsilon(t)$, defined by the solution of the equation

$$V^\varepsilon(t) = I + \int_0^t \Gamma(\kappa^\varepsilon(s/\varepsilon))V^\varepsilon(s)ds + \sum_{n=1}^{\nu^\varepsilon(t/\varepsilon)} [D^\varepsilon(\kappa_n^\varepsilon) - I]V^\varepsilon(\tau_n^\varepsilon -),$$

weakly convergent

$$V^\varepsilon(t) \Rightarrow \widehat{V}(t) \quad \text{as } \varepsilon \to 0,$$

to the averaged evolution $\widehat{V}(t)$ given by the solution of the equation

$$\widehat{V}(t) = I + \int_0^t [\widehat{\Gamma}(\hat{\kappa}(s)) + \widehat{D}(\hat{\kappa}(s))]\widehat{V}(s)ds,$$

or, in the evolutional equation form,

$$d\widehat{V}(t)/dt = [\widehat{\Gamma}(\hat{\kappa}(t)) + \widehat{D}(\hat{\kappa}(s))]\widehat{V}(t), \quad \widehat{V}(0) = I.$$

The proof of Theorem 4.3 is given in Section 4.8.

Remark 4.4. The averaged process in Theorem 4.3 is the stochastic evolution in the Markov random medium, but with a more simple, merged phase space $\widehat{X}$.

In various applications the merged space has only two states: $\widehat{X} = \{1,0\}$, that is meant that initial states of the switching process is splitting in two classes of states: $X = X_1 \cup X_0$ under the principle "will some given property be valid or not".

In the considered case the averaged evolution is constructed by a solution of two equations:

$$d\widehat{V}_k(t)/dt = [\widehat{\Gamma}(k) + \widehat{D}(k)]\widehat{V}_k(t), \quad k = 1,2.$$

The application of the splitting average algorithm determined by Theorem 4.3 is similar to that defined by Theorem 4.1 ergodic average algorithm.

Splitting diffusion approximation. Under some balance condition it is natural to consider a random evolution in diffusion approximation scheme with the splitting switching Markov medium.

Theorem 4.4. (*Splitting diffusion approximation*) Let the conditions C1 of Theorems 4.1–4.3 be valid and, in addition, the following condition be fulfilled:

C1: The balance condition is true on $\mathcal{B}_0$

$$\int_{X_k} \pi_k(dx)[\Gamma(x) + q(x)D(x)]\varphi = 0.$$

C2: The operators (4.38) are closed with the common dense domain $\mathcal{B}_0 \subseteq \mathcal{B}$ and the averaged operators

$$L(k) = \int_{X_k} \pi_k(dx)L(x), \quad k \in \widehat{X},$$

are the generators of strictly continuous semi-groups.

Then the random evolution $V^\varepsilon(t)$, $t \geq 0$, defined by the solution of the equation

$$V^\varepsilon(t) = I + \int_0^t \Gamma(\kappa^\varepsilon(s/\varepsilon^2))V^\varepsilon(s)ds + \sum_{n=1}^{\nu^\varepsilon(t/\varepsilon^2)} [D^\varepsilon(\kappa_n^\varepsilon) - I]V^\varepsilon(\varepsilon\tau_n^\varepsilon-),$$

is weakly convergent:

$$V^\varepsilon(t) \Rightarrow \widehat{W}(t) \quad \text{as} \quad \varepsilon \to 0$$

to the averaged evolution $\widehat{W}(t)$, $t \geq 0$, defined by the solution of the martingale problem

$$\widehat{W}(t)\varphi - \varphi - \int_0^t L(\hat{\kappa}(s))\widehat{W}(s)\varphi ds = \mu_t$$

for the generators $L(k)$, $k \in \widehat{X}$, and the martingale μ_t, $t \geq 0$, with respect to the filtration

$$\mathcal{F}_t = \sigma\{\widehat{W}(s)\varphi; \ 0 \leq s \leq t\}, \quad t \geq 0.$$

The proof of Theorem 4.4 is given in Section 4.8.

A similar remark 4.4 about the averaged evolution $\widehat{W}(t)$ in Theorem 4.4 can be given.

In applications the operators $L(k)$, $k \in \widehat{X}$, indeed, are defined by diffusion processes (Section 4.6).

Double averaging. Naturally, the averaged evolution obtained by the splitting average algorithm can be considered as the initial subject for further averaging under the condition that the merged Markov process $\hat{\kappa}(t)$, $t \geq 0$, is ergodic with the stationary distribution $\hat{\pi} = (\pi_k;\ 1 \leq k \leq N)$. Moreover, the splitting average algorithm can be applied several times. The last averaged evolution is obtained by the ergodic average algorithm. After that the resulting averaged evolution takes place without influence of a random medium.

The following problem arises: how can we get the resulting averaged evolution if the initial evolution is considered in a splitting random medium?

Now we consider the *double average algorithm* for a random evolution in a splitting Markov medium. For simplicity we examine the dynamic system in the following form:

$$du^\varepsilon(t)/dt = C(u^\varepsilon(t), \kappa^\varepsilon(t/\varepsilon^2)), \quad u^\varepsilon(0) = u. \tag{4.41}$$

Turn to notice that the switching process in (4.41) is considered with the accelerated scale ε^2 but without any balance condition.

Theorem 4.5. (*Double averaging*) Under conditions of Theorem 4.3 the weak convergence

$$u^\varepsilon(t) \Rightarrow \hat{u}(t) \quad \text{as } \varepsilon \to 0$$

takes place. The double averaged dynamic system $\hat{u}(t)$ is determined by the solution of the following equation

$$d\hat{u}(t)/dt = \widehat{C}(\hat{u}(t)),$$

$$\hat{u}(0) = u,$$

where the velocity of evolution $\widehat{C}(u)$ is determined by the relations

$$\widehat{C}(u) = \sum_{k=1}^N \pi_k \widehat{C}_k(u),$$

$$\widehat{C}_k(u) = \int_{X_k} \pi_k(dx)C(u,x). \tag{4.42}$$

Notice that the proof of Theorem 4.5 is noteworthy notwithstanding the heuristic nature of formula (4.42) (Section 4.8). The double average algorithm is defined by the formula (4.42).

Remark 4.5. The similar averaged result can be obtained after successive application of the splitting and ergodic average algorithms.

4.6 Application of average and diffusion approximation algorithms

We start with the rather simple stochastic system which is given by the *stochastic integral functional* (Section 4.1.)

$$\eta(t) = \eta_0 + \int_0^t a(\kappa(s))ds. \tag{4.43}$$

According to the interpretation of the stochastic integral functional (4.43) as a random evolution given in Corollary 4.5, the corresponding generators of semi-group is determined as follows

$$\Gamma(x)\varphi(u) = a(x)d\varphi(u)/du.$$

Hence the averaged generator is calculated by the proposition 4.1 in the following form

$$\Gamma\varphi(u) = ad\varphi(u)/du,$$

$$a = \int_X \pi(dx)a(x).$$

Returning to the phase space of the trajectories of the stochastic integral functional we get the averaged functional in the form:

$$\hat{\eta} = \eta_0 + at.$$

The average approach of the stochastic integral functional is given by the relation

$$\frac{1}{T}\int_0^T a(\kappa(s))ds \simeq at \quad \text{as} \quad T \to \infty.$$

The stochastic storage process is defined as follows (Section 4.2)

$$u(t) = u_0 + \int_0^t C(u(s), \kappa(s))ds + \sum_{n=1}^{\nu(t)} a(\kappa_n). \qquad (4.44)$$

The corresponding random evolution is described by the generator

$$\Gamma(x)\varphi(u) = C(u, x)d\varphi(u)/du, \quad x \in X,$$

and by the shift operators

$$D(x)\varphi(u) = \varphi(u + a(x)), \quad x \in X.$$

In order to apply the ergodic average algorithm it is necessary to introduce some series parameter $\varepsilon > 0$ into the jump component of the stochastic storage process in the following way:

$$D^\varepsilon(x)\varphi(u) = \varphi(u + \varepsilon a(x)), \quad x \in X.$$

Then the asymptotic expansion in (4.33) has the following form

$$D^\varepsilon(x)\varphi(u) = \varphi(u) + \varepsilon a(x)d\varphi(u)du + \varepsilon\theta_\varepsilon(x)\varphi(u).$$

So the main part of the jump component has the representation

$$D(x)\varphi(u) = a(x)d\varphi(u)/du, \quad x \in X.$$

Now we can apply the ergodic average algorithm described in proposition 4.1 to the stochastic storage process (4.44).

The averaged random evolution is determined as a solution of the equation

$$d\widehat{V}(t) = (\Gamma + D)\widehat{V}(t),$$

$$\widehat{V}(0) = I,$$

where the averaged operator Γ and D are defined by the following formulas:

$$\Gamma\varphi(u) = \widehat{C}(u)d\varphi(u)/du,$$

$$\widehat{C}(u) = \int_X \pi(dx)C(u,x),$$

$$D\varphi(u) = \hat{a}d\varphi(u)/du,$$

$$\hat{a} = q\int_X \rho(dx)a(x).$$

Returning to the trajectories phase space, the averaged storage process is the solution of the next equation

$$d\hat{u}(t)/dt = \widehat{C}(\hat{u}(t)) + \hat{a}\hat{u}(t),$$

$$\hat{u}(0) = u_0.$$

The stochastic additive functional of a Markov process is given by the relation (Section 4.1)

$$\eta(t) = \eta_0 + \int_0^t \xi(s, \kappa(s))ds + \sum_{n=1}^{\nu(t)} \alpha(\kappa_n), \qquad (4.45)$$

where $\xi(t,x)$, $x \in X$ is the family of processes with stationary and independent increments with cumulants $\Gamma(\lambda; x)$, and $x \in X$, and $\alpha(x)$, $x \in X$, is the totality of random variables with given distribution functions $G_x(u)$, $x \in X$. As it was mentioned above (Section 4.1) the processes $\xi(t; x)$ and the random variables $\alpha(x)$ are jointly independent for all fixed finite set of $x \in X$.

According to results of Section 4.3 the corresponding random evolution is determined by the generators $\Gamma(x)$, $x \in X$, which are given by

$$\Gamma(x)\varphi(u) = \int_{-\infty}^{\infty} e^{i\lambda u}\Gamma(\lambda; x)\tilde{\varphi}(\lambda)d\lambda.$$

The jump operators G_x, $x \in X$, which correspond to jumps of the stochastic additive functional (4.45) act in the following way:

$$G_x\varphi(u) = \int_{-\infty}^{\infty} \varphi(u+v)G_x(dv).$$

In order to get the averaged jump component we have to consider the jumps with the small series parameter $\varepsilon > 0$ introducing by

$$G_x^\varepsilon\varphi(u) = \int_{-\infty}^{\infty} \varphi(u+\varepsilon v)dG_x(v).$$

In accordance with (4.33) we get the asymptotic expansion

$$G_x^\varepsilon \varphi(u) = \varphi(u) + \varepsilon g(x)d\varphi(u)/du + \varepsilon \theta_\varepsilon(x)\varphi(u),$$

where

$$g(x) := \int_{-\infty}^{\infty} v G_x(dv) = E\alpha(x)$$

are the mean values of jumps.

Now applying the ergodic average algorithm described in proposition 4.1, we get the next averaged operators

$$\Gamma\varphi(u) = \int_{-\infty}^{\infty} e^{i\lambda u} \Gamma(\lambda)\tilde\varphi(\lambda)d\lambda$$

$$\Gamma(\lambda) = \int_X \pi(dx)\Gamma(\lambda; x) \qquad (4.46)$$

and

$$D\varphi(u) = gd\varphi(u)/du,$$

$$g := q\int_X \varrho(dx)g(x). \qquad (4.47)$$

So the averaged generator (4.46) defined by the cumulant of the process with stationary and independent increments $\hat\xi(t)$:

$$E\exp[i\lambda\hat\xi(t)] = \exp[t\Gamma(\lambda)].$$

The averaged operator (4.47) defines the uniform drift with rate g.

The averaged stochastic additive functional is given by the relation

$$\hat\eta(t) = \eta_0 + \hat\xi(t) + gt.$$

Note, that the averaged functional is stochastic but has the more simple character.

Remark 4.6. There is another form of ergodic averaging of the stochastic additive functional (4.45) by the ergodic theorem [18]:

$$\varepsilon \int_0^{t/\varepsilon} \xi(ds, \kappa(s)) \Rightarrow ct \quad \text{as} \quad \varepsilon \to 0,$$

Chapter 4

where the constant c is given in form

$$c = \int_X \pi(dx)c(x), \quad c(x) := E\xi(1; x).$$

This result can be obtained by the refinement of the average Theorem 4.1 with the following asymptotic condition:

$$\Gamma^\varepsilon(x)\varphi = \Gamma_0(x)\varphi + \theta_\varepsilon(x)\varphi, \quad \text{as} \quad \varepsilon \to 0, \quad \varphi \in \mathcal{B}_0,$$

$$\|\theta_\varepsilon(x)\varphi\| \to 0 \quad \text{as} \quad \varepsilon \to 0.$$

Diffusion approximation in applications. At the beginning we consider the *integral functional* (Corollary 4..5.)

$$\eta^\varepsilon(t) = \varepsilon^{-1} \int_0^t a(\kappa(s/\varepsilon^2))ds, \quad t \geq 0. \tag{4.48}$$

The corresponding continuous random evolution, according to Lemma 4.5, is defined by the relation

$$V^\varepsilon(t)\varphi(u) := E[\varphi(u + \eta^\varepsilon(t))/\kappa(s/\varepsilon^2), \ 1 \leq s \leq t]$$

and is generated by the generators

$$\Gamma(x)\varphi(u) = a(x)d\varphi(u)/du.$$

The main balance condition (4.37) of Theorem 4.2. has the following form:

$$\int_X \pi(dx)a(x) = 0. \tag{4.49}$$

Now we calculate the generator L by formula (4.38)–(4.39) taking into account that in our case $\Gamma^\varepsilon(x) = \Gamma(x)$ and $D^\varepsilon(x) = I$. Therefore we obtain

$$\begin{aligned} L(x)\varphi(u) &= \Gamma(x)R_0\Gamma(x)\varphi(u) \\ &= a(x)\frac{d}{du}R_0a(x)\frac{d}{du}\varphi(u) \\ &= a(x)R_0a(x)d^2\varphi(u)/du^2. \end{aligned}$$

We get the averaged generator L according to formula (4.39) in the following form:

$$L\varphi(u) = \sigma^2 d^2\varphi(u)/du^2, \tag{4.50}$$

where
$$\sigma^2 := \int_X \pi(dx)a(x)R_0a(x).$$

The generator (4.50) defines the diffusion process with the variance $2\sigma^2$.

The application of Theorem 4.4 to the integral functional (4.48) means that weak convergence

$$\varepsilon^{-1}\int_0^t a(\kappa(s/\varepsilon^2))ds = \varepsilon\int_0^{t/\varepsilon^2} a(\kappa(s))ds$$
$$\Rightarrow w(t) \quad \text{as } \varepsilon \to 0, \tag{4.51}$$

takes place and $w(0) = 0$.

More applicable form of the weak convergence (4.51) we obtain putting $\varepsilon^2 = T^{-1}$, and $\varepsilon = 1/\sqrt{T}$:

$$\int_0^{Tt} a(\kappa(s))ds/\sqrt{T} \Rightarrow w(t) \quad \text{as} \quad T \to \infty.$$

The parameter T can be interpreted as the scale of observations.

Now we consider the *stochastic additive functional* in the following normalized form (cf. with (4.51)):

$$\eta^\varepsilon(t) = u + \varepsilon\int_0^{t/\varepsilon^2} \xi(ds, \kappa(s))ds. \tag{4.52}$$

The corresponding continuous random evolution according to Corollary 4.3 is determined by the generators

$$\Gamma(x)\varphi(u) = \int_{-\infty}^{\infty} e^{i\lambda u}\Gamma(\lambda, x)\tilde{\varphi}(\lambda)d\lambda.$$

The asymptotic expansion (4.36) in this case has the following form

$$\Gamma(\varepsilon\lambda, x) = \varepsilon i\lambda c(x) - \varepsilon^2\lambda^2 c_0(x)/2 + o(\varepsilon^2). \tag{4.53}$$

Here $c(x) := E\xi(1, x)$ and $c_0(x) := E\xi^2(1, x) - c^2(x)$ are the first two semi-invariants of the process $\xi(t, x)$.

The balance condition has the following form

$$\int_X \pi(dx)c(x) = 0. \tag{4.54}$$

Now we calculate the generators $L(x)$ by formula (4.38) taking into account (4.53) in the following form

$$
\begin{aligned}
L_0(x)\varphi(u) &= \Gamma_0(x)\varphi(u) \\
&= \frac{c_0(x)}{2}d^2\varphi(u)/du^2,
\end{aligned}
$$

$$
\begin{aligned}
L_1(x)\varphi(u) &= \Gamma(x)R_0\Gamma(x)\varphi(u) \\
&= c(x)\frac{d}{du}R_0c(x)\frac{d}{du}\varphi(u) \\
&= c(x)R_0c(x)d^2\varphi(u)/du^2.
\end{aligned}
$$

Therefore according to the formula (4.39) the generator L has the following form

$$
L\varphi(u) = \frac{\sigma^2}{2}\, d^2\varphi(u)/du^2, \tag{4.55}
$$

where

$$
\sigma^2 = \int_X \pi(dx)[c_0(x) + 2c(x)R_0c(x)]. \tag{4.56}
$$

The generator (4.55) defines the diffusion process $w(t)$, $t \geq 0$, with the variance (4.56) and zero drift. So by Theorem 4.2, the diffusion approximation of the additive functional (4.52) under the balance condition (4.54) is given by the relation

$$
\int_0^{Tt} \xi(ds, \kappa(s))ds/\sqrt{T} \Rightarrow w(t) \quad \text{as} \quad T \to \infty.
$$

A *stochastic storage process* is defined by the solution of the evolutional equation

$$
du^\varepsilon(t)/dt = C(u^\varepsilon(t), \kappa(t/\varepsilon^2)),
$$

$$
u^\varepsilon(0) = u.
$$

The corresponding random evolution is described by the generators

$$
\Gamma(x)\varphi(u) = C(u, x)d\varphi(u)/du.
$$

The balance condition (4.37) of Theorem 4.2 has the following form:

$$
\int_X \pi(dx)C(u, x) \equiv 0.
$$

Now we calculate the generator L by formula (4.38), taking into account that in our case $\Gamma^\varepsilon(x) = \Gamma(x)$ and $D^\varepsilon(x) = I$. We obtain

$$
\begin{aligned}
L(x)\varphi(u) &= \Gamma(x)R_0\Gamma(x)\varphi(u) \\
&= C(u,x)\frac{d}{du}R_0C(u,x)\frac{d}{du}\varphi(u) \\
&= C(u,x)R_0C(u,x)d^2\varphi(u)/du^2 + C(u,x)R_0C'_u(u,x)d\varphi(u)/du.
\end{aligned}
$$

With the help of formula (4.39) we get the averaged generator in the following form:

$$
L\varphi(u) = \frac{\sigma^2(u)}{2}d^2\varphi(u)/du^2 + b(u)d\varphi(u)/du \tag{4.57}
$$

where

$$
\sigma^2(u) = 2\int_X \pi(dx)C(u,x)R_0C(u,x),
$$

$$
b(u) = \int_X \pi(dx)C(u,x)R_0C'(u,x).
$$

The generator (4.57) defines the diffusion process $w(t)$, $t \geq 0$, which is given by the solution of the stochastic equation.

$$
dw(t) = b(w(t))dt + \sigma(w(t))d\omega_t
$$

where ω_t, $t \geq 0$, is a standard Wiener process.

4.7 Counting processes

Constructive representation of jump Markov and semi-Markov processes can be realized using the counting process

$$
\nu(t) := \max\{n > 0 : \tau_n \leq t\} \tag{4.58}
$$

where τ_n, $n \geq 0$ are Markov renewal moments of jumps (see Section 2.2 and Definition 2.6). Meanwhile, the counting process can be considered applying the fundamental approach of the theory of semimartingales [13]. In this section, counting processes are considered as some stochastic systems in random media, i.e., as random evolutions.

Let $\{\kappa_n, \theta_n, \ n \geq 0\}$ be a Markov renewal process taking values in $E \times [0, \infty)$. The imbedded Markov chain $(\kappa_n, n \geq 0)$ takes values in a Polish space $(E, \mathcal{E})$ [4]. By Definition 2.4 a Markov renewal process is defined by the semi-Markov kernel

$$Q(x, B, t) = \mathcal{P}\{\kappa_{n+1} \in B, \ \theta_{n+1} \leq t / \kappa_n = x\},$$

where $x \in E$, $B \in \mathcal{E}$, $t \geq 0$.

For simplicity, in what follows, we assume that the components κ_n and θ_n are conditionally independent (see Section 2.1.3), that means

$$Q(x, B, t) = P(x, B) G_x(t). \tag{4.59}$$

Nonnegative random variables θ_n, $n \geq 1$, are called *renewal times* defining the intervals between the Markov renewal moments

$$\tau_n := \sum_{k=1}^{n} \theta_k, \quad \theta_n = \tau_n - \tau_{n-1}, \quad n \geq 1, \quad \tau_0 = 0.$$

The Markov renewal moments together with the imbedded Markov chain $(\kappa_n, \tau_n; \ n \geq 0)$ are also said to be a Markov renewal process.

Definition 4.4. [17] An *integer-valued random measure* is defined by the relation

$$\mu(dx, dt) := \sum_{n \geq 1} \delta_{\kappa_n, \tau_n}(dx, dt) 1_{(\tau_n < \infty)} \tag{4.60}$$

where δ_a is the Dirac measure concentrated in point a.

Therefore for any non-negative continuous function $w(x, t)$

$$\int_0^t \int_E w(x, s) \mu(dx, ds) = \sum_{n=1}^{\nu(t)} w(\kappa_n, \tau_n).$$

Note that the integer-valued random measure (4.59) can be considered as a *counting measure* for the Markov renewal process $(\kappa_n, \theta_n; \ n \geq 0)$ and is said to be a *multivariate point process* [17, Definition III.1.23].

By Theorem III.1.26 [17], there exists the unique *predictable random measure* $\bar{\mu}(dx, dt)$ which is a *compensator* of the multivariate point process μ, that is

$$\int_0^t \int_E w(x, s) [\mu(dx, ds) - \bar{\mu}(dx, ds)]$$

is a local martingale [17].

Moreover, it is possible to give the constructive representation for the compensator. Introduce the conditional distributions of the Markov renewal process

$$
\begin{aligned}
G_n(dx, dt) &:= \mathcal{P}\{\kappa_{n+1} \in dx,\ \tau_{n+1} \in dt / \mathcal{F}_n\} \\
&= \mathcal{P}\{\kappa_{n+1} \in dx, \tau_{n+1} \in dt / \kappa_n, \tau_n\} \\
&= Q(\kappa_n, dx, dt - \tau_n) \\
&= P(\kappa_n, dx) G_{\kappa_n}(dt - \tau_n).
\end{aligned}
$$

Particularly,

$$
\begin{aligned}
\bar{H}_n(t) &:= H_n([t, +\infty)) \\
&:= 1 - G_{\kappa_n}(t - \tau_n) \\
&= \bar{G}_{\kappa_n}(t - \tau_n).
\end{aligned}
$$

By Theorem III.1.33, [17], the compensator of the multivariate point process (4.59) can be represented as follows:

$$
\bar{\mu}(dx, dt) = \sum_{n \geq 0} 1_{(\tau_n < t \leq \tau_{n+1})} G_n(dx, dt) / \bar{H}_n(t). \tag{4.61}
$$

Therefore the compensator of the counting process (4.58) is represented by the relation

$$
\bar{\nu}(t) = \sum_{n \geq 0} \int_{\tau_n}^{t \wedge \tau_{n+1}} H_n(ds) / \bar{H}_n(s). \tag{4.62}
$$

Let the family of distributions $G_x(t)$, $x \in E$, be represented in the form

$$
\bar{G}_x(t) = \exp[-\lambda(x, t)], \quad \Lambda(x, t) = \int_0^t \lambda(x, s) ds. \tag{4.63}
$$

Lemma 4.7. The compensator (4.62) of the counting process (4.58) for the semi-Markov process with absolutely continuous distributions (4.63) is represented in the following form

$$
\bar{\nu}(t) = \int_0^t \lambda(\kappa(s), \gamma(s-)) ds \tag{4.64}
$$

where
$$\gamma(s) := s - \tau(s),$$
is the defect process,
$$\tau(s) := \tau_{\nu(t)}$$
is the point process.

Proof of Lemma 4.7. Taking into account (4.61) and (4.63) we calculate for $\tau_{n+1} < t$

$$
\begin{aligned}
\int_{\tau_n}^{\tau_{n+1}} H_n(ds)/\bar{H}_n(s) &= \int_{\tau_n}^{\tau_{n+1}} G_{\kappa_n}(ds - \tau_n)/\bar{G}_{\kappa_n}(s - \tau_n) \\
&= \int_0^{\theta_{n+1}} G_{\kappa_n}(ds)/\bar{G}_{\kappa_n}(s) \\
&= \int_0^{\theta_{n+1}} \lambda(\kappa_n, s)ds \\
&= \Lambda(\kappa_n, \theta_{n+1})
\end{aligned}
$$

and, similarly, for $\tau_{n+2} \geq t > \tau_n$

$$
\begin{aligned}
\int_{\tau_n}^{t} H_n(ds)/\bar{H}_n(s) &= \Lambda(\kappa_n, t - \tau_n) \\
&= \Lambda(\kappa_n, \gamma(t)).
\end{aligned}
$$

Hence

$$
\begin{aligned}
\bar{\nu}(t) &= \sum_{n \geq 0} \int_{\tau_n}^{t \wedge \tau_{n+1}} H_n(ds)/\bar{H}_n(s) \\
&= \sum_{n=0}^{\nu(t)} \Lambda(\kappa_n, \theta_{n+1}) + \Lambda(\kappa(t), \gamma(t)) \\
&= \int_0^t \lambda(\kappa(s), \gamma(s - 1))ds.
\end{aligned}
$$

Corollary 4.8. The compensator of the counting process for the Markov jump process with the intensity function $q(x)$, $x \in X$, is represented in the following form:

$$\bar{\nu}(t) = \int_0^t q(\kappa(s))ds. \tag{4.65}$$

That is the compensator (4.65) is an integral functional (Section 4.1) of the Markov jump process $\kappa(s)$, $s \geq 0$.

Now, the average and diffusion approximation algorithms can be applied basing on limit Theorems 4.1 and 4.2 concerning the compensator $\bar{\nu}(t)$ and the counting process $\nu(t)$ as well.

Corollary 4.9. Let a Markov jump process $\kappa(t)$, $t \geq 0$ given on a separable metric space $(X, \mathcal{X})$ with uniformly bounded intensity function

$$\sup_{x \in X} q(x) \leq C < +\infty \qquad (4.66)$$

be uniformly ergodic with the stationary distribution $\pi(dx)$.

Then the weak convergence as $\varepsilon \to 0$ take place:

$$\varepsilon\bar{\nu}(t/\varepsilon) = \varepsilon \int_0^{t/\varepsilon} q(\kappa(s))ds \Rightarrow tq = t \int_X \pi(dx)q(x) \qquad (4.67)$$

and

$$\varepsilon\nu(t/\varepsilon) \Rightarrow \nu(t), \qquad (4.68)$$

where $\nu(t)$, $t \geq 0$, is a Poisson process with intensity $tq = E\nu(t)$.

Proof. The first convergence follows from Theorem 4.1. To prove (4.68), note ([13]) that the martingales

$$\mu_t^\varepsilon = \varepsilon[\nu(t/\varepsilon) - \bar{\nu}(t/\varepsilon)] \qquad (4.69)$$

has the square characteristic

$$\langle \mu^\varepsilon \rangle_t = \varepsilon\bar{\nu}(t/\varepsilon) = \int_0^t q(\kappa(s/\varepsilon))ds. \qquad (4.70)$$

It follows from condition (4.66) that the compensators and, therefore, martingales (4.69) are compact. So there exists $\varepsilon_n \to 0$ that the weak convergence in (4.68) takes place:

$$\mu_t = \nu(t) - tq. \qquad (4.71)$$

The unique solution of the martingale problem for a Poisson process means that t he weak convergence (4.68) takes place.

Corollary 4.10. Let the semi-Markov process $\kappa(t)$, $t \geq 0$, on a separable metric space $(X, \mathcal{X})$ be defined by semi-Markov kernel (4.59). The uniformly ergodic imbedded Markov chain κ_n, $n \geq 0$, has the stationary distribution $\rho(dx)$. The family of distributions $G_x(t)$, $x \in X$, admit the representation (7) with uniform in $x \in X$ bounded intensities

$$\sup_{x \in X} \lambda(x, s) \leq \lambda(s) < +\infty. \tag{4.72}$$

Then the following weak convergences as $\varepsilon \to 0$ take place:

$$\varepsilon \int_0^{t/\varepsilon} \lambda(\kappa(s), \gamma(s-))ds \Rightarrow tm^{-1}, \tag{4.73}$$

$$m := \int_X \rho(dx)m(x), \quad m(x) := \int_0^\infty \bar{G}_x(t)dt, \tag{4.74}$$

and

$$\varepsilon \nu(t/\varepsilon) \Rightarrow \nu(t), \tag{4.75}$$

where $\nu(t)$, $t \geq 0$, is the Poisson process with the intensity

$$tm^{-1} = E\nu(t).$$

Proof. The existence of weak limits in (4.73) and (4.75) is proved similarly to the proof of Corollary 4.9. Note that the Markov process $\kappa(t)$, $\gamma(t)$, $t \geq 0$, has the following stationary distribution [12]:

$$\pi(dx, ds) = \rho(dx)\bar{G}_x(s)ds/m. \tag{4.76}$$

The limit constant in (4.73) is represented as follows

$$q = \int_X \rho(dx) \int_0^\infty \bar{G}_x(s)ds \, \lambda(x, s)/m. \tag{4.77}$$

Taking into account (4.63) we calculate

$$\begin{aligned}
\int_0^\infty \bar{G}_x(s)\lambda(x, s)ds &= -\int_0^\infty d_s\bar{G}_x(s) \\
&= \bar{G}_x(0) \\
&= 1.
\end{aligned}$$

Hence, in (4.77), we obtain

$$q = m^{-1}.$$

4.8 Proofs of limit theorems

The average limit theorem given in Section 4.4 and Section 4.5 are proved using the solutions of singular perturbed problems for reducible-invertible operators (Section 3.2). The starting idea is that the martingale characterization of Markov evolution has the following form (Section 4.3)

$$V(t)\varphi(\kappa(t)) - \int_0^t LV(s)\varphi(\kappa(s))ds = \mu_t,$$

where μ_t, $\mathcal{F}_t$, $t \geq 0$, is martingale and $\mathcal{F}_t = \sigma\{\kappa(s),\ 0 \leq s \leq t\}$.

As it was mentioned in Section 4.3 the random evolution has the interpretation as some stochastic evolution is described by the two-component Markov process. Therefore without loss of generality we can consider a stochastic interpretation of the random evolution $V(t)$, $t \geq 0$, in the following form

$$V(t)\varphi(u, x) = E[\varphi(u + \zeta(t), \kappa(t))/\kappa(s),\ 0 \leq s \leq t,\ \kappa(0) = x]$$

where two component Markov process $\zeta(t), \kappa(t)$, $t \geq 0$, is defined by the generator

$$L\varphi(u, x) = [Q + \Gamma(x) + Q_0[D(x) - I]]\,\varphi(u, x).$$

In the series scheme with small series parameter $\varepsilon > 0$ the random evolution can be represented as follows

$$V^\varepsilon(t)\varphi(u, x) = E\left[\varphi(u + \zeta^\varepsilon(t), \kappa^\varepsilon(t))/\kappa^\varepsilon(s),\ 0 \leq s \leq t/\varepsilon,\ \kappa^\varepsilon(0) = x\right].$$

The weak convergence of random evolutions

$$V^\varepsilon(t) \Rightarrow \hat{V}(t) \quad \text{as} \quad \varepsilon \to 0,$$

considered in Section 4.4 and Section 4.5 means that the weak convergence

$$\zeta^\varepsilon(t) \Rightarrow \hat{\zeta}(t) \quad \text{as } \varepsilon \to 0,$$

takes place [4].

We will prove Theorems 4.1–4.5 using Theorem 3.4 Section 3.4 and the martingale characterization of two-component Markov process $\zeta^\varepsilon(t)$, $\kappa^\varepsilon(t)$, $t \geq 0$.

The scheme of the proof of the average limit theorem for the random evolution is the same for the different theorems.

But there are distinctions between algorithmic steps of proof under various assumptions of dependence on the series parameter $\varepsilon > 0$.

So that considering the proof of limit theorems together we have the opportunity to restrict all the details of proof to only one of the limit theorems.

Proof of Theorem 4.1.   (*Ergodic average*)   The random evolution in Theorem 4.1 has the following form (Section 4.3):

$$V^\varepsilon(t) = I + \int_0^t \Gamma(\kappa(s/\varepsilon))V^\varepsilon(s)ds + \sum_{n=1}^{\nu(t/\varepsilon)} [D^\varepsilon(\kappa_n^\varepsilon) - I]V^\varepsilon(\tau_n^\varepsilon -).$$

Therefore the martingale characterization of such evolution has the following form (Section 4.3):

$$V^\varepsilon(t)\varphi(\kappa(t/\varepsilon)) - \int_0^t L^\varepsilon V^\varepsilon(s)\varphi(\kappa(s/\varepsilon))ds = \mu_t^\varepsilon, \qquad (4.78)$$

where the generator L^ε has the representation

$$L^\varepsilon\varphi = [\varepsilon^{-1}Q + \Gamma(x) + \varepsilon^{-1}Q_0[D^\varepsilon(x) - I]]\varphi, \quad \varphi \in \mathcal{B}_0.$$

Under condition C3 of the Theorem 4.1 the generator L^ε can be represented as an asymptotic expansion

$$L^\varepsilon\varphi = [\varepsilon^{-1}Q + L + Q_0\theta_\varepsilon]\varphi,$$

where

$$L := \Gamma(x) + Q_0 D(x) \qquad (4.79)$$

and

$$\|\theta_\varepsilon\varphi\| \to 0 \quad \text{as} \quad \varepsilon \to 0.$$

At the first step of proof of Theorem 4.1 we choose the test-function $\varphi^\varepsilon(x)$ which is dependent on ε in the following manner [16]

$$\varphi^\varepsilon(x) = \varphi + \varepsilon\varphi_1(x).$$

Then we can use Lemma 3.2 in order to get the solution of the *singular perturbed problem*

$$[\varepsilon^{-1}Q + L][\varphi + \varepsilon\varphi_1(x)] = \psi + \varepsilon\psi_{1\varepsilon}(x),$$

which is realized by the vector determined by the equalities

$$\hat{L}\varphi = \psi,$$

$$\varphi_1(x) = R_0(\psi - Q_1\varphi),$$

$$\psi_{1\varepsilon}(x) = Q_1 R_0(\psi - Q_1\varphi).$$

The averaged operator $\hat{L}$ is determined by the relation

$$\Pi L \Pi = \hat{L}\Pi, \qquad (4.80)$$

where the projector Π by the condition C2 of Theorem 4.1 acts in the following way

$$\Pi\varphi(x) = \int_X \pi(dx)\varphi(x)$$
$$=: \hat{\varphi}$$

where $\pi(dx)$ is the stationary distribution of the Markov process $\kappa(t)$, $t \geq 0$.

We are reminded that there is the following connection between stationary distributions of the Markov process $\kappa(t)$, $t \geq 0$, and of the imbedded Markov chain κ_n, $n \geq 0$:

$$\pi(dx)q(x) = q\rho(dx),$$

$$q = \int_X \pi(dx)q(x).$$

On the second step of the proof we calculate the averaged operator $\hat{L}$ using the relations (4.79) and (4.80)

$$\begin{aligned}
\Pi L \Pi\varphi(x) &= \Pi[\Gamma(x) + Q_0 D(x)]\hat{\varphi} \\
&= \int_X \pi(dx)[\Gamma(x) + q(x)\Pi D(x)]\hat{\varphi} \\
&= \hat{\Gamma}\hat{\varphi} + q\int_X \rho(dx)D(x)\hat{\varphi} \\
&= [\hat{\Gamma} + \hat{D}]\hat{\varphi}.
\end{aligned}$$

Therefore the algorithmic step of proof gives us the averaged operator

$$\hat{L} = \hat{\Gamma} + \hat{D}.$$

Now the martingale characterization of the random evolution in the ergodic average scheme has the following form

$$V^\varepsilon(t)\varphi - \int_0^t \widehat{L}V^\varepsilon(s)\varphi ds = \mu_t^\varepsilon + \eta_t^\varepsilon,$$

where the stochastic process

$$\eta_t^\varepsilon = \int_0^t [Q_0\theta_\varepsilon\varphi^\varepsilon + \varepsilon\psi_{1\varepsilon}]ds - \varepsilon V^\varepsilon(t)\varphi_1,$$

taking into account the conditions C3 and C4 of Theorem 4.1, satisfies the following condition (see (3.63) of Theorem 3.4, Section 3.4)

$$\sup_{\varepsilon\leq\varepsilon_0} E \sup_{0\leq t\leq T} \|\eta_t^\varepsilon\| \to 0 \quad \text{as} \quad \varepsilon_0 \to 0. \qquad (4.81)$$

On the third step of proof we calculate the square characteristic of the martingale (4.78). It is not so difficult to realize that

$$\langle\mu^\varepsilon\rangle_t = \int_0^t \psi^\varepsilon(s)ds, \qquad (4.82)$$

where the random process $\psi^\varepsilon(s), s \geq 0$, satisfies the following condition for every fixed $T > 0$ and some $\delta > 0$ (see (3.62))

$$E \sup_{0\leq t\leq T} \|\psi^\varepsilon(s)\|^{1+\delta} \to 0 \quad \text{as} \quad \varepsilon \to 0. \qquad (4.83)$$

Therefore, under conditions (4.81) and (4.83) by Theorem 3.4 (Section 3.4) we conclude that the weak convergence

$$V^\varepsilon(t)\varphi \Rightarrow \widehat{V}(t)\varphi \quad \text{as} \quad \varepsilon \to 0 \qquad (4.84)$$

takes place where the limiting evolution $\widehat{V}(t)$ is the solution of the following equation

$$\widehat{V}(t) - \int_0^t \widehat{L}\widehat{V}(s)ds = 0$$

or, in equivalent differential form

$$d\widehat{V}(t)/dt = [\widehat{\Gamma} + \widehat{D}]\widehat{V}(t).$$

At the last step we consider an applied interpretation of random evolutions in the following form (Section 4.3)

$$V^\varepsilon(t)\varphi(u) = E[\varphi(u + \zeta^\varepsilon(t))/\kappa(s/\varepsilon), 0 \le s \le t]$$

and hence the weak convergence (4.64) means that there exists the process $\hat\zeta(t)$ such that the weak convergence

$$\zeta^\varepsilon(t) \Rightarrow \hat\zeta(t) \quad \text{as } \varepsilon \to 0,$$

takes place. The proof of Theorem 4.1. is completed.

Proof of Theorem 4.2. (*Ergodic diffusion approximation*) The random evolution in Theorem 4.2 has the following form (Section 4.3)

$$V^\varepsilon(t) = I + \int_0^t \Gamma^\varepsilon(\kappa(s/\varepsilon^2))V^\varepsilon(s)ds + \sum_{n=1}^{\nu(t/\varepsilon^2)} [D^\varepsilon(\kappa_n^\varepsilon) - I]V^\varepsilon(\varepsilon\tau_n^\varepsilon -)$$

where the generator $\Gamma^\varepsilon(x)$ and a bounded linear operators $D^\varepsilon(x), x \in X$ from the condition C1 of Theorem 4.2 admit the following asymptotic expansions on $\varphi \in \mathcal{B}_0$

$$D^\varepsilon(x)\varphi = [I + \varepsilon D(x) + \varepsilon^2 D_0(x) + \varepsilon^2 \theta_{\varepsilon_1}(x)]\varphi, \tag{4.85}$$

$$\Gamma^\varepsilon(x)\varphi = [\Gamma(x) + \varepsilon\Gamma_0(x) + \varepsilon\theta_{\varepsilon_2}(x)]\varphi \tag{4.86}$$

and by the balance condition C2

$$\int_X \pi(dx)[\Gamma(x) + q(x)D(x)]\varphi = 0,$$

$$\varphi \in \mathcal{B}_0.$$

The martingale characterization of the random evolution $V^\varepsilon(t)$ provides the generator

$$L^\varepsilon\varphi = [\varepsilon^{-2}Q + \varepsilon^{-1}\Gamma^\varepsilon(x) + \varepsilon^{-2}Q_0[D^\varepsilon(x) - I]]\varphi.$$

Under condition (4.85) and (4.86) we get the following asymptotic expansion

$$L^\varepsilon\varphi = [\varepsilon^{-2}Q + \varepsilon^{-1}L_1 + L_0 + \theta_\varepsilon(x)]$$

where

$$L_1(x) = \Gamma(x) + Q_0 D(x), \quad L_0(x) = \Gamma_0(x) + Q_0 D_0(x)$$

and

$$\theta_\varepsilon(x)\varphi = [\theta_{\varepsilon_1}(x) + Q_0\theta_{\varepsilon_2}(x)]\varphi, \quad \varphi \in \mathcal{B}_0.$$

Therefore

$$\|\theta_\varepsilon(x)\varphi\| \to 0 \quad \text{as } \varepsilon \to 0, \quad \varphi \in \mathcal{B}_0.$$

At the main step of the proof of Theorem 4.2 we use the Lemma 3.3 in order to get the solution of the singular perturbation problem for the generator L^ε in the following form

$$[\varepsilon^{-2}Q + \varepsilon^{-1}L_1 + L_0]\varphi^\varepsilon(x) = \psi + \varepsilon\psi_1^\varepsilon$$

where the test-function are chosen as follows

$$\varphi^\varepsilon(x) = \varphi + \varepsilon\varphi_1(x) + \varepsilon^2\varphi_2(x).$$

With Lemma 3.3 we obtain the following representation

$$L^\varepsilon\varphi^\varepsilon(x) = \psi + \varepsilon\psi_1^\varepsilon$$

which is realized by the vectors which are determined by the following relations

$$L\varphi = \psi,$$

$$\varphi_1(x) = -R_0 L_1 \varphi,$$

$$\varphi_2(x) = R_0(\psi - L_2\varphi),$$

$$\psi_1^\varepsilon = [L_1 + \varepsilon L_0]\varphi_2 + L_0\varphi_1.$$

The averaged operator L is determined by the relations

$$L(x) = L_0(x) - L_1(x)R_0 L_1(x),$$

$$L\Pi = \Pi L(x)\Pi.$$

Therefore the averaged operator L is determined in the form

$$L = \int_X \pi(dx)L(x).$$

Now the martingale characterization of the random evolution in the ergodic diffusion approximation scheme has the form

$$V^\varepsilon(t)\varphi - \int_0^t LV^\varepsilon(s)\varphi ds = \mu_t^\varepsilon + \eta_t^\varepsilon$$

where the stochastic process η_t^ε and the martingale μ_t^ε satisfy the conditions of the pattern limit Theorem 3.4 (Section 3.4).

Therefore we can conclude that the weak convergence

$$V^\varepsilon(t)\varphi \Rightarrow \widehat{W}(t)\varphi \quad \text{as} \quad \varepsilon \to 0,$$

takes place. The limiting evolution $\widehat{V}(t)$ is determined by the solution of the martingale problem for the generator $\widehat{L}$:

$$\widehat{W}(t)\varphi - \int_0^t L\widehat{W}(s)\varphi ds = \mu_t, \quad \varphi \in \mathcal{B}_0.$$

The proof of Theorem 4.2 is completed.

Proof of Theorem 4.3. (*Splitting average*) The random evolution in Theorem 4.3 is defined by the solution of the following equation

$$V^\varepsilon(t) = I + \int_0^t \Gamma(\kappa^\varepsilon(s/\varepsilon))V^\varepsilon(s)ds + \sum_{n=1}^{\nu^\varepsilon(t/\varepsilon)} [D^\varepsilon(\kappa_n^\varepsilon) - I]V^\varepsilon(\tau_n^\varepsilon-).$$

Therefore the martingale characterization can be presented as

$$V^\varepsilon(t)\varphi - \int_0^t L^\varepsilon V^\varepsilon(s)\varphi ds = \mu_t^\varepsilon$$

where the generator L^ε has the form

$$L^\varepsilon\varphi = [\varepsilon^{-1}Q^\varepsilon + \Gamma(x) + \varepsilon^{-1}Q_0^\varepsilon[D^\varepsilon(x) - I]]\varphi$$

where

$$Q^\varepsilon = Q + \varepsilon Q_1,$$

$$Q_0^\varepsilon = Q_0 + \varepsilon Q_{10}$$

according to supposition of Theorem 4.3.

Under the conditions of Theorem 4.3 the generator L^ε has the following asymptotic expansion

$$L^\varepsilon \varphi = [\varepsilon^{-1} Q + L(x) + \theta_\varepsilon(x)]\varphi$$

where

$$L(x) = Q_1 + \Gamma(x) + Q_0 D(x)$$

and the operator $\theta_\varepsilon(x)$ satisfy the relation

$$\|\theta_\varepsilon(x)\varphi\| \to 0 \quad \text{as} \quad \varepsilon \to 0, \quad \varphi \in \mathcal{B}_0.$$

The main part of the proof of Theorem 4.3 is the calculation of the averaged operator L which is determined by the relation

$$L\Pi = \Pi L(x)\Pi$$

where the projector Π acts in the following way (Section 4.5)

$$\Pi\varphi(x) = \sum_{k=1}^{N} \hat{\varphi}_k 1_k(x),$$

$$\hat{\varphi}_k = \int_{X_k} \pi_k(dx)\varphi(x).$$

It is easy to check that the averaged operator L is represented as follows

$$L\varphi(x) = [\widehat{Q} + \widehat{\Gamma} + \widehat{D}]\hat{\varphi}$$

where $\widehat{Q}$ is defined by the relation

$$\Pi Q_1 \Pi = \widehat{Q}\Pi$$

and the averaged operator $\widehat{\Gamma}$ and $\widehat{D}$ are determined by the relation

$$\widehat{\Gamma}(k) = \int_{X_k} \pi_k(dx)\Gamma(x),$$

$$\widehat{D}(k) = q_k \int_{X_k} \rho_k(dx)D(x).$$

Now, using the solution of the singular perturbation problem of the Lemma 3.2 for the operator L^ε, we obtain the test functions $\varphi^\varepsilon(x) = \varphi + \varepsilon\varphi_1(x)$

$$L^\varepsilon \varphi^\varepsilon(x) = L\varphi + \varepsilon\psi_1^\varepsilon(x)$$

where the vector $\psi_1^\varepsilon(x)$ is determined by the relations (Section 3.2)

$$\psi_1^\varepsilon(x) = L(x)R_0[\psi - L(x)\varphi],$$

$$\psi = L\varphi.$$

The conclusive step of the proof of Theorem 4.3 is similar to the one of Theorem 4.1.

Proof of Theorem 4.4 (*Splitting diffusion approximation*) is omitted because it is analogous to the proof of Theorem 4.2 with some additional details which are presented in the proof of Theorem 4.3.

Proof of Theorem 4.5. (*Double average*) Under conditions of Theorem 4.5 we consider the random evolution in the following form

$$V^\varepsilon(t)\varphi(u, x) := E_{u,x}[\varphi(u^\varepsilon(t), \kappa^\varepsilon(t/\epsilon^2))/\kappa^\varepsilon(s), \quad 0 \leq s \leq t/\epsilon^2]$$

where the conditional expectation $E_{u,x}$ means that

$$u^\varepsilon(0) = u,$$

$$\kappa^\varepsilon(0) = x.$$

The start point of proof is the martingale characterization of the two-component Markov process $u^\varepsilon(t)$, $\kappa^\varepsilon(t) := \kappa^\varepsilon(t/\varepsilon^2)$, $t \geq 0$, $\varepsilon > 0$,

$$\varphi(u^\varepsilon(t), \kappa^\varepsilon(t)) - \int_0^t L^\varepsilon \varphi(u^\varepsilon(s), \kappa^\varepsilon(s))ds = \mu_t^\varepsilon$$

with the generators

$$L^\varepsilon \varphi(u, x) = [\varepsilon^{-2}Q^\varepsilon + \Gamma(x)]\varphi(u, x)$$

where

$$Q^\varepsilon = Q + \varepsilon Q_1$$

and

$$\Gamma(x)\varphi(u) = c(u, x)d\varphi(u)/du.$$

The main part of the proof of Theorem 4.5 is the application of the solution of the singular perturbation problem for the generator L^ε according to the

Lemma 3.4. So we get the following representation on the test-function $\varphi^\varepsilon = \varphi + \varepsilon\varphi_1 + \varepsilon^2\varphi_2$:

$$
\begin{aligned}
L^\varepsilon\varphi^\varepsilon &= [\varepsilon^{-2}Q + \varepsilon Q_1 + \Gamma(x)](\varphi + \varepsilon\varphi_1 + \varepsilon^2\varphi_2) \\
&= \widehat{\widehat{\Gamma}}\varphi + \varepsilon\theta_\varepsilon,
\end{aligned}
$$

where the double averaged operator $\widehat{\widehat{\Gamma}}$ is determined by the Lemma 3.4 in the following relations

$$\widehat{\Gamma}\widehat{\Pi} = \widehat{\Pi}\widehat{\Gamma}\widehat{\Pi},$$

$$\widehat{\Gamma}\Pi = \Pi\Gamma\Pi.$$

Note that under the condition of Theorem 4.5, the contracted operator $\widehat{Q}_1$ defined by the relation

$$\widehat{Q}_1\Pi = \Pi Q_1\Pi$$

is a reducible–invertible (see Section 3.6). Under the supposition of Theorem 4.5 we, first of all, calculate

$$\widehat{\Gamma}\varphi(u) = \widehat{C}_k(u)d\varphi(u)/du$$

where

$$\widehat{C}_k(u) = \int_{X_k} \pi_k(dx)C(u,x), \quad k \in \widehat{X},$$

and, after that,

$$\widehat{\widehat{\Gamma}}\varphi(u) = \widehat{C}(u) = \sum_{k=1}^{N} \hat{\pi}_k\widehat{C}_k(u).$$

The conclusive step of the proof of Theorem 4.5 is similar to the one of Theorem 4.1.

Chapter 5

Diffusion approximation of Markov queueing systems and networks

Markov processes with locally independent increments on the Euclidean phase space R^N are considered as mathematical models of Markovian queueing systems and networks. Average and diffusion approximation schemes with some small series parameter $\varepsilon > 0$ are developed and applied for some concrete queueing systems and networks.

5.1 Algorithms of diffusion approximation

The contemporary theory of stochastic systems has a wide variety of methods of analysis based on ideas of the simplified description of mathematical models. Algorithms of *diffusion approximation* (ADA) occu py a special place among these methods [1, 4, 6, 9]. The simplifying effect of such algorithms is based on the fact that all the variety of original parameters characterizing the evolution of a stochastic system, transforms into three functions which determine a diffusion process: centering shift, drift diffusion and variance. In such a way, there reaches an essential simplification of mathematical analysis of a diffusion model.

At the same time, in certain conditions of sufficiently high intensities of state transitions of the original system, the diffusion approximation presents a simplifying variant of system evolution with a practically sufficient degree of precision.

Among numerous ADA, the most perspective ones seem to be those algorithms for which the original evolution is described by a Markov jump process in finite dimensional Euclidean state space. For such processes there were developed effective ADA with corresponding software realization.

The efficiency of such algorithms is based on the fact that, the basic characteristics of diffusion approximation: centering shift, drift and variance, has enough simple expression through the basic parameters which define evolution of the original system.

The software realization also contains routines which simulate trajectories of Markov processes and which calculate stationary characteristics. Those routines permit a user to convince, on concrete examples, an efficiency of diffusion approximation.

The basic type of diffusion processes approximating a wide class of ergodic Markov jump processes, are Ornstein–Uhlenbeck ones which stationary distribution simply expresses through the normal distribution function. An essential particularity of ADA consists in a real possibility application to problems of optimization and control which are well-developed for diffusion processes.

A perspective property of ADA is also a possibility of their simple extension to stochastic systems with input flows (queueing requests etc.) described by Markov renewal processes. [10]

The most effective application of ADA seems to be for stochastic systems which describe queueing systems and networks as well as storage and transport processes widely used in problems of communication, insurance and various networks (computers, transport, industrial, biological etc.)

ADA, in perspective, can admit modifications, either in the direction of an extension of domain of their applications, or in the direction of use of wider class of diffusion processes, in particular those described by additional conditions in type of boundary conditions (absorbing, reflecting or transparent screens).

ADA are based on limit theorems which establish the validity of algorithms in the conditions of high intensity of functions of the system. In such a way, there eliminates arbitrariness of an heuristic approach for simplifying analysis of stochastic systems.

The efficiency of ADA easily checks the conditions which, for the first view, absolutely do not correspond to the forming of the ADA mathematical basis limit theorems.

For example, the classical supply energy system consisting of a finite number of devices working independently with every device consuming or producing energy with given intensities, can be effectively approximated by a diffusion process, even when the number of such devices is less then ten. One can be convinced of this fact in the course of ADA's demonstrative application.

Notice briefly the main classes of stochastic systems which admit diffusion approximation.

1. One-linear and multilinear queueing systems with equal or different devices with intensities depending on the current state of the system.

2. Queueing networks consisting of a finite number of nodes with equal or different systems in the nodes.

3. Storage systems describing by evolution differential equations with regard for influence of a Markov random environment to the velocities of storage or consumption.

4. General stochastic systems in Markov random environment having a property of stochastic determinism: the stochastic characteristics of evolution depend only on the current state of the system.

5.2 Markov queueing processes

Markov processes $\zeta(t)$, $t \geq 0$, with locally independent increments [2] in Euclidean phase space R^N can be determined by a generator L on a Banach space $\mathcal{B}_{R^N}$ of real-valued bounded measurable functions $\varphi(u)$, $u \in R^N$, with sup-norm $\|\varphi(u)\| := \sup_{u \in R^N} |\varphi(u)|$

$$L\varphi(u) = \int_{R^N} Q(u, dz)[\varphi(u + z) - \varphi(u)]$$

$$+C(u)\varphi'(u) + \frac{1}{2} \operatorname{Tr}[B(u)\varphi''(u)]. \tag{5.1}$$

The kernel $Q(u, dz)$, defining intensities of values of jumps is supposed to be a positive bounded measure:

$$Q(u, R^N) =: q(u) \in \mathcal{B}_R.$$

The vector function $C(u) = (c_k(u), 1 \le k \le N)$ sets the velocity of the deterministic drift $\rho(t)$ which is defined by a solution of the evolutional equation

$$d\rho(t)/dt = C(\rho(t)). \qquad (5.2)$$

The second term in (5.1) means the scalar product of two vector functions:

$$C(u)\varphi'(u) := \sum_{k=1}^{N} c_k(u) d\varphi(u)/du_k.$$

The pure diffusion part of the Markov process is defined by the generator

$$L_0\varphi(u) := \frac{1}{2}\mathrm{Tr}\left[B(u)\varphi''(u)\right]$$

$$:= \frac{1}{2}\sum_{k,r=1}^{N} b_{kr}(u)d^2\varphi(u)/du_k du_r. \qquad (5.3)$$

The covariance matrix of diffusion is

$$B(u) := [b_{kr}(u);\ 1 \le k, r \le N].$$

The deterministic drift $\rho(t)$ can be considered as part of the diffusion process with the generator

$$L^0\varphi(u) := C(u)\varphi'(u) + \frac{1}{2}\mathrm{Tr}\left[B(u)\varphi''(u)\right]. \qquad (5.4)$$

The increments of the Markov process $\zeta(t)$ on a small time interval Δt can be represented as a sum

$$\Delta\zeta(t) := \zeta(t + \Delta t) - \zeta(t) \simeq$$

$$\simeq \Delta\nu(t) + \Delta\rho(t) + \Delta w(t),$$

where $\nu(t)$ is a Markov jump process with the generator

$$Q\varphi(u) := \int_{R^N} Q(u, dz)[\varphi(u + z) - \varphi(u)]. \qquad (5.5)$$

The diffusion process $w(t)$ is defined by the generator (5.3).

The homogeneous in time Markov jump process $\nu(t)$ which is defined by the generator (5.5) is considered as a starting point for the mathematical model of the queueing system.

Example 5.1. *Systems with bounded input* [5]. There are given N identical devices. Each device can be in two different states, namely, a *working state* 1 and a *repairing state* 0. The intensities of working and repairing times are λ and μ respectively. In Feller's book [5, Section 18.5] such a system is called the "supplying energy system". The corresponding Markov *queueing process* $\nu(t)$ can be defined as the number of repairing (or working) devices at the instant of time t. Therefore the phase space is $E_N = \{n : 0 \le n \le N\}$ containing $N + 1$ states.

The intensities of jumps for *repairing devices* $\nu(t)$ are defined as follows:

$$Q(n, +1) = (N - n)\lambda, \quad 0 \le n \le N,$$
$$Q(n, -1) = n\mu, \quad 0 \le n \le N. \tag{5.6}$$

Besides

$$
\begin{aligned}
q(n) \; &:= \; Q(n, +1) + Q(n, -1) \\
&= \; (N - n)\lambda + n\mu, \quad 0 \le n \le N.
\end{aligned}
$$

Notice that

$$q(0) = N\lambda = Q(0, +1),$$
$$q(N) = N\mu = Q(N, -1).$$

These two equalities express the reflecting property of boundary points $n = 0, N$. The queueing process $\nu(t)$ in the phase space E_N determines a *Markovian random walk* on the finite state space E_N.

It is easy to realize [5, Section 17.7] that the stationary distribution of Markovian process $\nu(t)$

$$p_n := \mathcal{P}\{\nu(t) = n\} = \binom{N}{n} p^n q^{N-n}, \quad 0 \le n \le N,$$

exists with $p = \lambda/(\lambda + \mu)$, $q = 1 - p = \mu/(\lambda + \mu)$. Moreover the explicit expression for the distribution of the process $\nu(t)$:

$$\mathcal{P}\{\nu(t) = n\} = \binom{N}{n} A^n (1 - A)^{N-n},$$

where $A = p(1 - e^{-(\lambda+\mu)t})$, can be given [5, Problem 10.17, Chapter XVII].

This example is a rare occurrence of an explicit solution of the problem. Meanwhile, a system with bounded input can be considered in the series scheme with the natural series parameter $N \to \infty$.

Two approximation problems (averaging and diffusion approximations) can be considered similarly to the random evolutions considered in Chapter 4.

In order to get the certain average effect we need to choose an adequate normalized scheme which has the following form in the case of the classical averaging

$$\nu_N(t) := \nu(Nt)/N,$$

or, using a small series parameter $\varepsilon = 1/N$ the normalized queueing process in the average scheme is represented as follows:

$$\nu^\varepsilon(t) := \varepsilon\nu(t/\varepsilon). \tag{5.7}$$

Note that the normalized queueing process $\nu^\varepsilon(t)$ takes values in the phase space $E_\varepsilon = \{u = \varepsilon n,\ 0 \le n \le N\}$.

Considered queueing process in the series scheme as $N \to \infty$ (or, $\varepsilon \to 0$) is dependent on the series parameter N but, for the sake of simplicity, we suppose this dependence without marking it.

The generator of the normalized queueing process (5.7) has the following form:

$$Q^\varepsilon\varphi(u) = \varepsilon^{-1} \int_R Q^\varepsilon(u, dz)[\varphi(u + \varepsilon z) - \varphi(u)] \tag{5.8}$$

where the intensity kernel $Q^\varepsilon(u, dz)$ with jumps in points $z = \pm 1$, is defined by (5.6)

$$\begin{aligned} Q^\varepsilon(u, +1) &= (1 - u)\lambda, \quad u = \varepsilon n, \quad 0 \le n \le N, \\ Q^\varepsilon(u, -1) &= u\mu, \quad u = \varepsilon n, \quad 0 \le n \le N. \end{aligned} \tag{5.9}$$

Considering this generator on a continuously differentiable function $\varphi(u)$ we obtain the following asymptotic representation

$$\begin{aligned} Q^\varepsilon\varphi(u) &\simeq \int_R zQ^\varepsilon(u, dz)\varphi'(u) \\ &= [\lambda - (\lambda + \mu)u]\varphi'(u) \end{aligned}$$

the generator

$$\mathbf{C}\,\varphi(u) = [\lambda(1-u) - \mu u]\varphi'(u)$$

defines the limiting component $\rho(t)$ which is determined by a solution of the evolutional equation

$$d\rho(t)/dt = C(\rho(t)), \quad C(u) := \lambda - (\lambda+\mu)u.$$

The solution of this equation is represented as follows:

$$\rho(t) = p + (\rho_0 - p)e^{-(\lambda+\mu)t},$$

where $\rho_0 := \rho(0) = \mathcal{P}\lim_{\varepsilon\to 0}\nu^\varepsilon(0)$ or it can be represented in another form

$$\nu_N(0) = N\rho_0 + o(N).$$

Hence, choosing $\rho_0 = p = \lambda/(\lambda+\mu)$ we obtain the equilibrium point for the normalized queueing process:

$$\mathcal{P}\lim_{N\to\infty}\nu(Nt)/N = \lambda/(\lambda+\mu).$$

Another approximation problem is to investigate the fluctuations of the queueing process around the equilibrium point. The fluctuations have the following normalized classical form:

$$\zeta_N(t) := [\nu(Nt)/N - p]\sqrt{N}.$$

By choosing a small series parameter $\varepsilon = 1/\sqrt{N}$ we obtain the following normalized form for fluctuations:

$$\zeta^\varepsilon(t) := \varepsilon\nu(t/\varepsilon^2) - \varepsilon^{-1}p.$$

The generator of this Markov process is determined by the relation

$$L^\varepsilon\varphi(u) = \varepsilon^{-2}\int_R Q^\varepsilon(p+\varepsilon u, dz)[\varphi(u+\varepsilon z) - \varphi(u)] \qquad (5.10)$$

where $p = \lambda/(\lambda+\mu)$ and the kernel $Q^\varepsilon(u,dz)$ is defined by (5.9).

Considering the generator (5.10) on the set of twice continuously differentiable functions $\varphi(u)$ we can obtain the following asymptotic representation:

$$L^\varepsilon \varphi(u) \simeq \varepsilon^{-2}[\varepsilon \int_R zQ^\varepsilon(p + \varepsilon u, dz)\varphi'(u) +$$
$$+\frac{\varepsilon^2}{2}\int_R z^2 Q^\varepsilon(p + \varepsilon u, dz)\varphi''(u)].$$

Taking into account (5.9) and the fact that $C(p) = 0$ this approximation is transformed to

$$L^\varepsilon \varphi(u) \simeq L^0 \varphi(u)$$

where

$$L^0 \varphi(u) = -(\lambda + \mu)u\varphi'(u) + B^2 \varphi''(u),$$

$$B^2 := \lambda\mu/(\lambda + \mu).$$

This generator defines the Ornstein–Uhlenbeck diffusion process $\zeta^0(t)$ which can be considered as a diffusion approximation of the normalized queueing process $\zeta^\varepsilon(t)$ or, in another form,

$$[\nu(Nt)/N - p]\sqrt{N} \simeq \zeta^0(t).$$

For applications, we can use the following approximation of the queueing process:

$$\nu(Nt) \simeq Np + \zeta^0(t)\sqrt{N}.$$

The schemes of averaging and diffusion approximation of the Markov queueing process considered before can be extended on a rather wide class of Markov processes. The heuristic form of the before mentioned reasoning will be verified in the next section.

5.3 Average and diffusion approximation

Examples considered in Section 5.2 have common tools in average and diffusion approximation schemes for the rather wide class of Markovian processes.

5.3.1 Average scheme

The queueing process in the average normalized scheme is considered in the following form:

$$\nu^\varepsilon(t) := \varepsilon \nu^\varepsilon(t/\varepsilon). \tag{5.11}$$

That is the generator of the Markov process $\nu^\varepsilon(t)$ has the following representation:

$$Q^\varepsilon \varphi(u) = \varepsilon^{-1} \int_{R^N} Q^\varepsilon(u, dz)[\varphi(u + \varepsilon z) - \varphi(u)]. \tag{5.12}$$

The main condition in the average approximation scheme is the asymptotical representation of the first moment of jumps.

Theorem 5.1. (*Averaging*) Let the approximation condition hold:

$$C^\varepsilon(u) := \int_{R^N} z Q^\varepsilon(u, dz) = C(u) + \theta^\varepsilon(u) \tag{5.13}$$

with the negligible residual term

$$\|\theta^\varepsilon(u)\| \to 0 \quad \text{as } \varepsilon \to 0; \tag{5.14}$$

and let a positive solution of the equation

$$d\rho/dt = C(\rho(t)), \quad \rho(0) = \rho_0 \geq 0, \tag{5.15}$$

exists. Then the normalized queueing process (5.11) converges in distribution as $\varepsilon \to 0$ to a solution $\rho(t)$ of the equation (5.15) with the initial value $\rho_0 = \mathcal{P}\lim_{\varepsilon \to 0} \varepsilon \nu^\varepsilon(0)$:

$$\varepsilon \nu^\varepsilon(t/\varepsilon) \Rightarrow \rho(t) \quad \text{as } \varepsilon \to 0. \tag{5.16}$$

Remark 5.1. Let the initial value $\rho_0 \geq 0$ be an equilibrium point for the velocity $C(u)$:

$$C(\rho_0) = 0.$$

Then the queueing process in the average normalized scheme has the equilibrium point as well:

$$\varepsilon \nu^\varepsilon(t/\varepsilon) \Rightarrow \rho_0 \quad \text{as } \varepsilon \to 0,$$

uniformly on every finite time interval $t \in [0, T]$.

Proof of Theorem 5.1. The scheme of proof considered in Section 5.2 is used. Under the approximation condition (5.13) the generator (5.12) acting on the continuously differentiable function $\varphi(u)$ has the following asymptotic representation

$$Q^\varepsilon \varphi(u) = \mathbf{C}\,\varphi(u) + \theta^\varepsilon(u), \tag{5.17}$$

where the generator

$$\mathbf{C}\,\varphi(u) = C(u)\varphi'(u) \tag{5.18}$$

determines the evolutional equation (5.15). The residual term $\theta^\varepsilon(u)$ in (5.17) (different from the term in (5.13)) satisfies the negligible condition (5.14). Now using the martingale characterization of a Markov process (see Section 3.3)

$$\mu_t^\varepsilon = \varphi(\nu^\varepsilon(t)) - \int_0^t Q^\varepsilon \varphi(\nu^\varepsilon(s))\, ds \tag{5.19}$$

and the asymptotical representation (5.17) we get the following asymptotical form for the martingale (5.19)

$$\mu_t^\varepsilon = \varphi(\nu^\varepsilon(t)) - \int_0^t \mathbf{C}\,\varphi(\nu^\varepsilon(s))ds + \psi_t^\varepsilon$$

where the residual term ψ_t^ε satisfies the condition

$$E[\sup_{0 \le t \le T} |\psi_t^\varepsilon|] \to 0 \quad \text{as } \varepsilon \to 0. \tag{5.20}$$

It is easy to verify that the square characteristic of martingale (5.19) admits the representation (3.61) with a random function $\zeta^\varepsilon(s)$ satisfying condition (3.65). Hence, by Theorem 3.4 the weak convergence (5.16) takes place. The limiting process $\rho(t)$ is defined by a solution of the equation

$$d\varphi(\rho(t))/dt = \mathbf{C}\,\varphi(\rho(t)) \tag{5.21}$$

which is equivalent to the evolutional equation (5.15) for the limiting process $\rho(t)$.

5.3.2 Diffusion approximation scheme

A Markov queueing process in the diffusion approximation scheme is considered in the following centered and normalized form:

$$\zeta^\varepsilon(t) := \varepsilon\nu^\varepsilon(t/\varepsilon^2) - \varepsilon^{-1}\rho^\varepsilon(t) \qquad (5.22)$$

where the Markov jump process $\nu^\varepsilon(t)$ is determined by the generator

$$Q^\varepsilon\varphi(u) = \int_{R^N} Q^\varepsilon(u, dz)[\varphi(u+z) - \varphi(u)]. \qquad (5.23)$$

The centered function $\rho^\varepsilon(t)$ is defined by a solution of the evolutional equation

$$d\rho^\varepsilon(t)/dt = C^\varepsilon(\rho^\varepsilon(t)). \qquad (5.24)$$

The more general centering scheme with the velocity $C^\varepsilon(u)$ depending on a series parameter ε is motivated by the real queueing system with bounded input considered in Section 5.6. In spite of the fact that the Markov jump process $\nu^\varepsilon(t)$ is homogeneous in time, the normalized centered process (5.22) is heterogeneous but in a special way due to the deterministic centered function $\rho^\varepsilon(t)$. Meanwhile, the two component and homogeneous in time Markov process $\zeta^\varepsilon(t), \rho^\varepsilon(t)$, $t \geq 0$, can be used in the diffusion approximation scheme.

Lemma 5.1. The generator of the coupled Markov process $\zeta^\varepsilon(t)$, $\rho^\varepsilon(t)$, $t \geq 0$, has the following representation

$$L^\varepsilon\varphi(u, v) = L_0^\varepsilon\varphi(u, v) + C^\varepsilon(v)\varphi_v'(u, v) \qquad (5.25)$$

where

$$L_0^\varepsilon\varphi(u, v) := \varepsilon^{-2}Q^\varepsilon(v)\varphi(u, v) - \varepsilon^{-1}C^\varepsilon(v)\varphi_u'(u, v) \qquad (5.26)$$

and the generator $Q^\varepsilon(v)$ acts in the following way

$$Q^\varepsilon(v)\varphi(u) := \int_{R^N} Q^\varepsilon(v + \varepsilon u, dz)[\varphi(u + \varepsilon z) - \varphi(u)]. \qquad (5.27)$$

Proof of Lemma 5.1. The increments of the first component have the following representation:

$$\begin{aligned} \Delta\zeta^\varepsilon(t) &:= \zeta^\varepsilon(t+\Delta) - \zeta^\varepsilon(t) \\ &= \varepsilon\Delta\nu^\varepsilon(t/\varepsilon^2) - \varepsilon^{-1}\Delta\rho^\varepsilon(t). \end{aligned}$$

According to equation (5.24)

$$\Delta \rho^\varepsilon(t) := \rho^\varepsilon(t + \Delta) - \rho^\varepsilon(t)$$
$$= C^\varepsilon(\rho^\varepsilon(t))\Delta + o(\Delta).$$

Hence

$$\Delta \zeta^\varepsilon(t) = \varepsilon \Delta \nu^\varepsilon(t/\varepsilon^2) - \varepsilon^{-1} C^\varepsilon(\rho^\varepsilon(t))\Delta + o(\Delta).$$

Now the conditional expectation

$$E_{u,v}\varphi(\zeta^\varepsilon(t + \Delta), \rho^\varepsilon(t + \Delta)) :=$$

$$E[\varphi(\zeta^\varepsilon(t + \Delta), \rho^\varepsilon(t + \Delta))/\zeta^\varepsilon(t) = u, \rho^\varepsilon(t) = v]$$

can be estimated as follows:

$$E_{u,v}\varphi(\zeta^\varepsilon(t + \Delta), \rho^\varepsilon(t + \Delta)) =$$

$$= E_u\varphi(u + \varepsilon \Delta \nu^\varepsilon(t/\varepsilon^2), v) + C^\varepsilon(v)[\varphi'_v(u,v) - \varphi'_u(u,v)]\Delta + o(\Delta). \quad (5.28)$$

Note that, under condition $\zeta^\varepsilon(t) = u$, $\rho^\varepsilon(t) = v$, according to the definition of the normalized centered process (5.22),

$$\varepsilon^2 \nu^\varepsilon(t/\varepsilon^2) = v + \varepsilon u.$$

Therefore, the first term in (5.28) can be estimated as follows:

$$E_u\varphi(u + \varepsilon \Delta \nu^\varepsilon(t/\varepsilon^2), v) =$$

$$= E[\varphi(u + \varepsilon \Delta \nu^\varepsilon(t/\varepsilon^2), v)/\varepsilon^2 \nu^\varepsilon(t/\varepsilon^2) = v + \varepsilon u] =$$

$$= \varphi(u,v) + \varepsilon^{-2} Q^\varepsilon(v)\varphi(u,v)\Delta + o(\Delta) \quad (5.29)$$

with the generator $Q^\varepsilon(v)$ defined by (5.27). The relations (5.28) and (5.29) imply (5.25)–(5.27).

Note that the intensity kernel $Q^\varepsilon(u, dz)$ defines the intensity of jump values for the process $\nu^\varepsilon(t)$ under condition

$$\varepsilon^2 \nu^\varepsilon(t) = u$$

which is fixed. The diffusion approximation for the normalized centered queueing process (5.22) is realized under asymptotical representation for the infinitesimal characteristics defined by the generator (5.25)–(5.27).

Theorem 5.2. (*Diffusion approximation*) Let the intensity of jump values of the Markov queueing process $\nu^\varepsilon(t)$ admit the following asymptotical representation of the first and the second moments as $\varepsilon \to 0$:

$$b^\varepsilon(v, u) := \int_{R^N} z Q^\varepsilon(v + \varepsilon u, dz) =$$

$$= b(v) + \varepsilon b^1(v, u) + \varepsilon \theta^\varepsilon(v, u), \tag{5.30}$$

$$B^\varepsilon(v, u) := \int^{R^N} z z^* Q^\varepsilon(v + \varepsilon u, dz) =$$

$$= B(v) + \theta^\varepsilon(v, u) \tag{5.31}$$

where residual terms $\theta^\varepsilon(v, u)$ in both formulas, being different, satisfy the negligible condition

$$\|\theta^\varepsilon(v, u)\| \to 0 \quad \text{as } \varepsilon \to 0. \tag{5.32}$$

Let the velocity of the centered function $C^\varepsilon(v)$ admit the asymptotical representation

$$C^\varepsilon(v) = C(v) + \varepsilon C^1(v) + \varepsilon \theta^\varepsilon(v) \tag{5.33}$$

with the negligible term $\theta^\varepsilon(v)$:

$$\|\theta^\varepsilon(v)\| \to 0 \quad \text{as } \varepsilon \to 0.$$

Then, under the additional balance condition

$$b(v) = C(v) \tag{5.34}$$

the normalized centered queueing process $\zeta^\varepsilon(t)$, defined by (5.22), converges in distribution to a diffusion process $\zeta^0(t)$:

$$\zeta^\varepsilon(t) \Rightarrow \zeta^0(t) \quad \text{as } \varepsilon \to 0.$$

The generator of the limiting diffusion is then defined by

$$L_t^0 \varphi(u) = [b^1(\rho(t), u) - C^1(\rho(t))] \varphi'(u) + \frac{1}{2} \text{Tr} \, [B(\rho(t)) \varphi''(u)] \tag{5.35}$$

where the centered function $\rho(t)$ is defined by a solution of the evolutional equation

$$d\rho(t)/dt = C(\rho(t)) \tag{5.36}$$

with the initial value

$$\rho_0 = \mathcal{P} \lim_{\varepsilon \to 0} \varepsilon^2 \nu^\varepsilon(0). \tag{5.37}$$

The initial value of the limit diffusion is defined by the condition

$$\zeta^0(0) = \mathcal{P} \lim_{\varepsilon \to 0} \zeta^\varepsilon(0).$$

Remark 5.2. Let the intensity kernel have the following asymptotical representation:

$$Q^\varepsilon(v, dz) = Q^0(v, dz) + \varepsilon Q^1(v, dz) \tag{5.38}$$

and the kernel $Q^0(v, dz)$ is differentiable with respect to v. Then the asymptotical representation (5.30) has the following terms:

$$
\begin{aligned}
b(v) &= \int_{R^N} z Q^0(v, dz), \\
b^1(v, u) &= b^1(v) + u b'(v), \\
b^1(v) &= \int_{R^N} z Q^1(v, dz).
\end{aligned}
\tag{5.3.39}
$$

Remark 5.3. Let the velocity $C(v)$ of the centered function have an equilibrium point $\rho_0 \geq 0$: $C(\rho_0) = 0$. Then under initial condition (5.37) and the additional condition (5.38) the limiting diffusion of the Ornstein–Uhlenbeck type is defined by the generator

$$L_0^0 \varphi(u) = a(u) \varphi'(u) + \frac{1}{2} \mathrm{Tr}\, [B \varphi''(u)], \tag{5.40}$$

where the drift function is linear with respect to u:

$$a(u) = a_0 + a_1 u,$$

$$a_0 = b^1(\rho_0), \quad a_1 = b'(\rho_0), \tag{5.41}$$

and the covariance matrix is constant:

$$B = B(\rho_0). \tag{5.42}$$

Proof of Theorem 5.2. The asymptotic representation of the generator (5.25)–(5.27) is considered on the set $\mathcal{D}$ of real-valued twice continuously differentiable functions $\varphi(u, v)$ with a compact support.

Taking into account the asymptotical representations (5.30) and (5.31) of the first two moments of jump values and (5.33) for the velocity of the centered function we obtain

$$
\begin{aligned}
L_0^\varepsilon \varphi(u, v) &:= \varepsilon^{-2} Q^\varepsilon(v) \varphi(u, v) - \varepsilon^{-1} C^\varepsilon(v) \varphi'_u(u, v) \\
&= L_0^0 \varphi(u, v) + \varepsilon^{-1} [b(v) - C(v)] \varphi'_u(u, v) + \theta^\varepsilon(v, u)
\end{aligned}
$$

$$
L_0^0 \varphi(u, v) = [b^1(v, u) - C^1(v)] \varphi'_u(u, v) + \frac{1}{2} \mathrm{Tr}\ [B(v) \varphi''_u(u, v)]. \qquad (5.43)
$$

As before, the residual term $\theta^\varepsilon(v, u)$ satisfies the negligible condition (5.32). The balance condition (5.34) yields the following asymptotical representation for the generator (5.25)

$$
L^\varepsilon \varphi(u, v) = L_0^0 \varphi(u, v) + C(v) \varphi'_v(u, v) + \theta^\varepsilon(u, v).
$$

Now the martingale characterization of the coupled Markov process $\zeta^\varepsilon(t)$, $\rho^\varepsilon(t)$, $t \geq 0$,

$$
\mu_t^\varepsilon = \varphi(\zeta^\varepsilon(t), \rho^\varepsilon(t)) - \int_0^t L^\varepsilon \varphi(\zeta^\varepsilon(s), \rho^\varepsilon(s)) ds \qquad (5.44)
$$

can be transformed into

$$
\begin{aligned}
\mu_t^\varepsilon = \ & \varphi(\zeta^\varepsilon(t), \rho^\varepsilon(t)) \\
& - \int_0^t [L_0^0 \varphi(\zeta^\varepsilon(s), \rho^\varepsilon(s)) + C(\rho^\varepsilon(s))\, \varphi'_v(\zeta^\varepsilon(s), \rho^\varepsilon(s))] ds + \psi_t^\varepsilon
\end{aligned}
$$

with the residual negligible term satisfying the condition (3.63) of Theorem 3.4. More complicated calculations yield to the representation of the square characteristic of the martingale (5.44) in the form (3.61) with a random function $\zeta^\varepsilon(s)$ satisfying the compactness condition (3.62). Hence, by Theorem 3.4, the weak convergence takes place

$$
(\zeta^\varepsilon(t), \rho^\varepsilon(t)) \Rightarrow (\zeta^0(t), \rho(t)) \quad \text{as } \varepsilon \to 0.
$$

The limiting two-component Markov process $\zeta^0(t), \rho(t), t \geq 0$, is determined by the generator

$$
L^0 \varphi(u, v) = L_0^0 \varphi(u, v) + C(v) \varphi'_v(u, v). \qquad (5.45)
$$

To complete the proof of Theorem 5.2, note that the coupled Markov process $\zeta^0(t)$, $\rho(t)$, $t \geq 0$, with the first component defined by the generator (5.35), is defined by the generator (5.43), (5.45).

Remark 5.4. Using the centered drift function $\rho^\varepsilon(t)$ dependent on the series parameter ε, the drift term in (5.35) which is not dependent on u can be removed. But it is not possible to remove the drift part of limiting diffusion completely.

Remark 5.5. The considered class of Markov processes contains "density dependent population processes" for which in the book [4] the diffusion approximation without a drift coefficient was investigated. The limiting diffusion has not ergodicity in spite of the fact that the initial process is ergodic (see Example 5.1).

5.4 Stationary distribution

The diffusion approximation of a stationary Markov jump process has to provide the approximation of its stationary distribution. It means that the stationary distribution $\pi^\varepsilon(du)$, $\varepsilon > 0$, of normalized and centered queueing processes weakly converge as $\varepsilon \to 0$ to the stationary distribution $\pi^0(du)$ of the limiting diffusion process.

Essential aspects of this problem are considered in the book [4, Section 4.9]. The main idea is to verify relatively compactness of stationary distributions π^ε, $\varepsilon > 0$. The most useful approach involves construction of the Lyapunov function $V(u)$ satisfying the following conditions

$V(u) > 0$ in some vicinity of zero point $u = 0$;

$V(0) = 0$;

$V(u) \to \infty$ as $\|u\| \to \infty$.

In addition, $V(u)$ is an *excessive* function for the generator L_0^0 of the limiting diffusion process:

$$L_0^0 V(u) \leq 0.$$

For a given diffusion process the Ito formula provides a more direct approach [4, Section 4.9, Problem 35].

A weak convergence of stationary distributions is considered for the normalized queueing process with an equilibrium point $\rho \geq 0$:

$$\zeta^\varepsilon(t) = \varepsilon \nu^\varepsilon(t/\varepsilon^2) - \varepsilon^{-1}\rho. \tag{5.46}$$

The generator of the Markov process $\zeta^\varepsilon(t)$ has the following form:

$$L_0^\varepsilon \varphi(u) = \varepsilon^{-2} Q^\varepsilon(\rho)\varphi(u) \tag{5.47}$$

where

$$Q^\varepsilon(\rho)\varphi(u) = \int_{R^N} Q^\varepsilon(\rho + \varepsilon u, dz)[\varphi(u + \varepsilon z) - \varphi(u)]. \tag{5.48}$$

The limiting diffusion process in this case, under conditions of Theorem 5.2 is determined by the generator

$$L_0^0 \varphi(u) = a(u)\varphi'(u) + \frac{1}{2}\mathrm{Tr}\,[B\varphi''u] \tag{5.49}$$

where

$$a(u) = b^1(\rho, u) \quad \text{and} \quad B = B(\rho). \tag{5.50}$$

Theorem 5.3. (*Convergence of stationary distributions*) Consider the Lyapunov function $V(u)$ for the generator (5.49) satisfying condition

$$L_0^0 V(u) \leq C < 0. \tag{5.51}$$

Then the stationary distributions π^ε, $\varepsilon > 0$, of the normalized queueing process (5.46) defined by the generators (5.47) converge weakly to the stationary distribution π^0 of the limiting diffusion process, defined by the generator (5.49).

Proof of Theorem 5.3. At first, we establish the weak compactness of the stationary distributions π^ε, $\varepsilon > 0$. By the definition of the generator we have representation

$$E_u V(\zeta^\varepsilon(t + h)) = V(\zeta^\varepsilon(t)) + h L_0^\varepsilon V(\zeta^\varepsilon(t)) + o(h). \tag{5.52}$$

Under conditions of Theorem 5.2. we obtain the following asymptotical representation

$$L_0^\varepsilon V(u) = L_0^0 V(u) + \theta^\varepsilon(u) \tag{5.53}$$

with the negligible term $\theta^\varepsilon(u)$ satisfying the condition (5.14).

Putting (5.51)–(5.53) together we get the following inequality

$$E_u V(\zeta^\varepsilon(t+h)) \le V(\zeta^\varepsilon(t)) + h[C + o_h(\varepsilon)]$$

where $o_h(\varepsilon) \to 0$ as $\varepsilon \to 0$. Hence, for a smooth enough h and ε the source inequality

$$E_n V(\zeta^\varepsilon(t+h)) \le V(\zeta^\varepsilon(t)) \qquad (5.54)$$

takes place.

It follows that $V(\zeta^\varepsilon(t))$, $t \ge 0$, is a supermartingale:

$$E_u V(\zeta^\varepsilon(t)) \le E_u V(\zeta^\varepsilon(0)) = V(u)$$

for all $t > 0$.

Let $R_k^N := \{u : \|u\| > k\}$ and $V_k := \inf_{u \in R_k^N} V(u)$. Due to the property of the Lyapunov function

$$V_k \to \infty \quad \text{as } k \to \infty.$$

Now the weak compactness of the stationary distributions π^ε, $\varepsilon > 0$, is the consequence of the estimate obtained with the help of Chebyshev's inequality for the transition probabilities:

$$P_t^\varepsilon(u, R_k^N) := \mathcal{P}\{\zeta^\varepsilon(t) \in R_k^N / \zeta^\varepsilon(0) = u\} \le$$

$$\le V(u)/V_k \to 0 \quad \text{as } k \to \infty$$

for all small enough $\varepsilon > 0$.

But by the property of stationary distribution

$$\pi^\varepsilon(R_k^N) = \lim_{t \to \infty} P_t^\varepsilon(u, R_N^N) \le V(u)/V_k \to 0 \quad \text{as } k \to \infty.$$

Therefore, there exists a convergent sequence

$$\lim_{\varepsilon_n \to 0} \int_{R^N} \pi^{\varepsilon_n}(du)\varphi(u) = \int_{R^N} \pi^0(du)\varphi(u). \qquad (5.55)$$

We have to establish now that the measure $\pi^0(du)$ is a stationary distribution of the limiting diffusion process defined by the generator (5.49).

Let $P_t^0(u, A) := \mathcal{P}\{\zeta^0(t) \in A/\zeta^0(0) = u\}$ be the transition probabilities of the limiting diffusion process defined by the generator (5.49).

By Theorem 5.2 the convergence

$$\lim_{\varepsilon \to 0} \int_{R^N} \varphi(z) P_t^\varepsilon(u, dz) = \int_{R^N} \varphi(z) P_t^0(u, dz) \qquad (5.56)$$

takes place uniformly for $t \in [0, T]$ for any finite function $\varphi(u)$ with a compact support. Introduce the measure

$$\mu_T^\varepsilon(u, A) := \int_0^T P_t^\varepsilon(u, A) dt$$

and the functional

$$\mu_T^\varepsilon(u, \varphi) := \int_{R^N} \varphi(z) \mu_T^\varepsilon(u, dz).$$

Convergence (5.56) provides the convergence

$$\lim_{\varepsilon \to 0} \mu_T^\varepsilon(u, \varphi) = \mu_T^0(u, \varphi) = \int_{R^N} \varphi(z) \int_0^T P_t^0(u, dz) dt. \qquad (5.57)$$

Now we calculate

$$\int_{R^N} \pi^\varepsilon(du) \mu_T^\varepsilon(u, \varphi) = T \int_{R^N} \pi^\varepsilon(du) \varphi(u). \qquad (5.58)$$

Using the relative compactness of π^ε, $\varepsilon > 0$ established above, we get by (5.55) and (5.57)

$$\lim_{\varepsilon_n \to 0} \int_{R^N} \pi^{\varepsilon_n}(du) \mu_T^{\varepsilon_n}(u, \varphi) = \int_{R^N} \pi^0(du) \mu_T^0(u, \varphi)$$

and by (5.58)

$$\lim_{\varepsilon_n \to 0} \int_{R^N} \pi^{\varepsilon_n}(du) \mu_T^{\varepsilon_n}(u, \varphi) = T \int_{R^N} \pi^0(du) \varphi(u).$$

Putting the last two limits together we obtain

$$\int_{R^N} \pi^0(du) \int_{R^N} \varphi(z) \int_0^T P_t^0(u, dz) dt = T \int_{R^N} \pi^0(dz) \varphi(z).$$

It follows that

$$\pi^0(dz) = \frac{1}{T} \int_0^T dt \int_{R^N} \pi^0(du) P_t^0(u, dz).$$

Hence $\pi^0(dz)$ is the stationary measure of the limiting diffusion process defined by the limiting generator (5.49). The proof of Theorem 5.3 is completed.

The construction of the Lyapunov function $V(u)$ can also be realized in another way.

  Let us consider the limiting diffusion process defined by the generator (5.40)–(5.42) with the constant covariance matrix and the linear drift function with respect to u. Such a limiting diffusion process has a wide scope of applications (see Section 5.5).

Lemma 5.2. Let a diffusion process on the real line R have the generator

$$L_0^0\varphi(u) = a(u)\varphi'(u) + \frac{1}{2}B\varphi''(u) \qquad (5.59)$$

where the drift function

$$a(u) = a_0 - a_1 u, \quad a_1 > 0. \qquad (5.60)$$

Then the Lyapunov function for the generator (5.59) can be chosen as follows:

$$V(u) = V_0 \int_0^u e^{b(z)}dz[V_1 - \int_0^z e^{-b(y)}dy], \qquad (5.61)$$

where the constant $V_0 > 0$ and V_1 satisfies inequality

$$V_1 > \int_0^\infty e^{-b(y)}dy. \qquad (5.62)$$

Here

$$b(z) := -2\int_0^z a(u)du/B^2. \qquad (5.63)$$

Proof of Lemma 5.2. Note that the integral in the right-hand side of (5.62) is finite due to the assumption (5.60) and (5.63). Using the first and the second derivative of the function (5.61) it is easy to calculate

$$L_0^0 V(u) = -V_0 B^2/2 < 0.$$

Hence, the function $V(u)$ given by (5.61) is excessive for the generator (5.59).

5.5 Markovian queueing systems

There are two possibilities to construct the diffusion approximation for real stochastic systems. Either to apply directly the pattern limit Theorem 5.2 to a real queueing system and, then, compute the local characteristics of the limiting diffusion process by formulas (5.39), or, to formulate some intermediate algorithm for a certain class of queueing systems. The second way is chosen.

5.5.1 Collective limit theorem in R^1

Consider a normalized queueing process in such a form

$$\zeta^\varepsilon(t) = \varepsilon \nu^\varepsilon(t/\varepsilon^2) - \varepsilon^{-1}\rho(t) \tag{5.64}$$

where the queueing process $\nu^\varepsilon(t)$ determines the number of customers in the system including that which have to be served at time t. The centering function $\rho(t)$ is some deterministic positive valued continuously differentiable. For simplicity assume that customers arrive and leave the system one by one.

The basic assumption is that the intensities of input and service times under condition $\nu^\varepsilon(t) = k$ depend on k in the following way:

$$\lambda_\varepsilon(k) = \lambda(\varepsilon^2 k) + \varepsilon \lambda_1(\varepsilon^2 k),$$

$$\mu^\varepsilon(k) = \mu(\varepsilon^2 k) + \varepsilon \mu_1(\varepsilon^2 k) \tag{5.65}$$

with some given functions $\lambda(v)$, $\mu(v)$, $\lambda_1(v)$ and $\mu_1(v)$, $v \in R^1$.

Note that the connection between variables $k = \nu^\varepsilon(t)$, $v = \rho(t)$ and $u = \zeta^\varepsilon(t)$ is represented as follows:

$$\varepsilon^2 k = v + \varepsilon u.$$

Therefore, the intensities of jumps of Markov process (5.64) under condition $\zeta^\varepsilon(t) = u$ have the following representation

$$Q^\varepsilon(v + \varepsilon u, +1) = \lambda(v + \varepsilon u) + \varepsilon \lambda_1(v + \varepsilon u)$$

$$Q^\varepsilon(v + \varepsilon u, -1) = \mu(v + \varepsilon u) + \varepsilon \mu_1(v + \varepsilon u). \tag{5.66}$$

Theorem 5.4. (*Collective limit theorem*) Let the intensities of input and service times of the queueing process $\nu^\varepsilon(t)$ be set by relation (5.65) with continuously differentiable functions $\lambda(u)$ and $\mu(u)$, having a bounded first derivative, and continuous functions $\lambda_1(u)$ and $\mu_1(u)$. Let there exist the positive solution of the evolutional equation

$$d\rho(t)/dt = C(\rho(t)), \tag{5.67}$$

$$C(v) := \lambda(v) - \mu(v) \tag{5.68}$$

with the initial condition $\rho(0) = \rho_0 \geq 0$. The initial values of the queueing process converges in probability in such a way

$$\zeta^\varepsilon(0) = \varepsilon\nu^\varepsilon(0) - \varepsilon^{-1}\rho_0 \Rightarrow \zeta^0 \quad \text{as } \varepsilon \to 0. \tag{5.69}$$

Then the normalized queueing process (5.64) converges weakly as $\varepsilon \to 0$ to the diffusion process $\zeta^0(t)$:

$$\varepsilon\nu^\varepsilon(t/\varepsilon^2) - \varepsilon^{-1}\rho(t) \Rightarrow \zeta^0(t) \quad \text{as } \varepsilon \to 0.$$

The generator of the limiting diffusion process $\zeta^0(t)$ is determined by the relation:

$$L_t^0\varphi(u) = a(t,u)\varphi'(u) + \frac{1}{2}B(\rho(t))\varphi''(u) \tag{5.70}$$

where

$$a(t,u) = uC'(\rho(t)) + b^1(\rho(t)), \tag{5.71}$$

$$b^1(v) = \lambda_1(v) - \mu_1(v),$$

$$B(v) = \lambda(v) + \mu(v). \tag{5.72}$$

The initial condition $\zeta^0(0) = \zeta^0$.

Proof of Theorem 5.4. First we calculate the local characteristics of the coupled Markov process $\zeta^\varepsilon(t)$, $\rho(t)$, $t \geq 0$, using the representation (5.66) of the intensities of jumps and the formulas (5.30) and (5.31). The intensities of jumps (5.66) is transformed as follows:

$$Q^\varepsilon(v + \varepsilon u, +1) = \lambda(v) + \varepsilon[u\lambda'(v) + \lambda_1(v)] + \varepsilon\theta^\varepsilon(v,u),$$

$$Q^\varepsilon(v + \varepsilon u, +1) = \mu(v) + \varepsilon[u\mu'(v) + \mu_1(v)] + \varepsilon\theta^\varepsilon(v, u)$$

with the residual negligible terms $\theta^\varepsilon(v, u)$ satisfying the condition (5.32). By formulas (5.30), (5.38) and (5.39) we obtain

$$b(v) = \lambda(v) - \mu(v),$$

$$b^1(v) = \lambda_1(v) - \mu_1(v).$$

By formula (5.31) we get

$$B(v) = \lambda(v) + \mu(v).$$

At last, by balance condition (5.34)

$$C(v) = b(v) = \lambda(v) - \mu(v).$$

We finish proof of Theorem 5.4 applying Theorem 5.2.

Remark 5.6. In the stationary regime with the equilibrium point ρ:

$$C(\rho) = \lambda(\rho) - \mu(\rho) = 0 \tag{5.73}$$

the limiting diffusion process is of the Ornstein–Uhlenbeck type with the generator

$$L^0\varphi(u) = a(u)\varphi'(u) + \frac{1}{2}B\varphi''(u)$$

where

$$a(u) = ua + a_1,$$

$$a = \lambda'(\rho) - \mu'(\rho),$$

$$a_1 = \lambda_1(\rho) - \mu_1(\rho),$$

$$B = B(\rho). \tag{5.74}$$

The *algorithm for the diffusion approximation of Markov queueing systems* is formulated in the *collective limit* Theorem 5.4. Under given intensities of input and service times represented in the form (5.65) the centered function $\rho(t)$ is determined by a solution of the evolutional equation (5.67) with the velocity function defined by (5.68). The generator of the limiting diffusion is defined by (5.70) with the drift function given by (5.71) and the covariance function given by (5.72). The initial value of the limiting diffusion is defined by (5.69).

5.5.2 Systems of M/M type

A single served system with an unbounded queue is defined by the intensities of input λ and service time μ of exponential distributions.

Corollary 5.1. Under condition of heavy traffic, $\lambda > \mu$, the normalized queueing process

$$\zeta^\varepsilon(t) = \varepsilon \nu^\varepsilon(t/\varepsilon^2) - \varepsilon^{-1}\rho(t) \qquad (5.75)$$

with the centering function

$$\rho(t) = \rho_0 + (\lambda - \mu)t,$$

where ρ_0 is determined under the following conditions

$$\varepsilon \nu^\varepsilon(0) - \varepsilon^{-1}\rho_0 \Rightarrow \zeta^0 \quad \text{as } \varepsilon \to 0, \qquad (5.76)$$

converges weakly to the diffusion process $\zeta^0(t)$ with zero mean and variance

$$B = \lambda + \mu.$$

Proof. It is easy to check that the queueing system of type $M/M/1/\infty$ is obtained by taking the intensity function

$$\lambda(v) = \lambda, \quad \mu(v) = \mu.$$

Hence, according to Theorem 5.4, the velocity of the centering function is

$$C(v) = \lambda - \mu.$$

Therefore, the solution of the equation (5.67) is

$$\rho(t) = \rho_0 + (\lambda - \mu)t.$$

Further, according to (5.71)and (5.72) we obtain

$$a(t, u) = 0, \quad B(v) = \lambda + \mu.$$

Corollary 5.2. The unbounded server system, with the rate of Poisson arrivals λ and intensity of exponential distributed service time μ has $M/M/\infty$ type. The normalized queueing process (5.75) with the centering function

$$\rho(t) = \lambda/\mu + (\rho_0 - \lambda/\mu)e^{-\mu t}, \qquad (5.77)$$

where ρ_0 is determined by initial condition (5.76) converges weakly as $\varepsilon \to 0$ to the diffusion process $\zeta^0(t)$ defined by the generator (5.70) with the drift function

$$a(t, u) = -u\mu \tag{5.78}$$

and variance

$$B(\rho(t)) = 2\lambda + (\rho_0\mu - \lambda)e^{-\mu t}. \tag{5.79}$$

Proof. Now it is also obvious that a system of $M/M/\infty$ type is determined by the intensity function $\lambda(v) = \lambda$ and $\mu(v) = \mu v$. Applying the collective limit Theorem 5.4 we obtain by (5.68)

$$C(v) = \lambda - v\mu;$$

Hence the centering function is (5.77) and by (5.71) and (5.72) we obtain the drift function (5.78) and the variance (5.79).

Note that the stationary regime is obtained by putting $\rho_0 = \lambda/\mu$, i.e. the initial condition has the form:

$$\nu^\varepsilon(0) = \varepsilon^{-2}\lambda/\mu + \varepsilon^{-1}\zeta^0 + o(\varepsilon^{-1}).$$

Alternatively, the variance

$$B(\rho(t)) \to 2\lambda \quad t \to \infty$$

with exponential rate.

5.5.3   Systems with bounded input

Such a system was considered in Section 5.1 as an illustrative example with a preliminary heuristic scheme of asymptotic analysis. Now we can apply Theorem 5.4 to get the rigorous diffusion approximation result. According to Example 5.1 the intensities of input and service times, under the condition that the number of repairing devices $\nu^\varepsilon(t) = k$, have the following representation ($\varepsilon^2 = 1/N$, $v = k/N$):

$$\lambda_\varepsilon(k) = (N - k)\lambda = N(1 - v)\lambda,$$

$$\mu_\varepsilon(k) = k\mu = Nv\mu.$$

Taking into account (5.65) we obtain that a system with bounded input is defined by the intensity function

$$\lambda(v) = (1 - v)\lambda, \quad \mu(v) = v\mu.$$

Now, applying the algorithm of diffusion approximation formulated in Theorem 5.4, we calculate by (5.68), (5.71) and (5.72)

$$C(v) = \lambda - v(\lambda + \mu),$$

$$a(t, u) = -u(\lambda + \mu),$$

$$B(v) = (1 - v)\lambda + v\mu.$$

The equation (5.67) has the solution

$$\rho(t) = p + (\rho_0 - p)e^{-(\lambda+\mu)t}$$

where $p := \lambda/(\lambda + \mu)$ and the constant $\rho_0 \geq 0$ is defined by the initial condition

$$\rho_0 = \varepsilon^2 v^\varepsilon(0) - \varepsilon\zeta^0 + o(\varepsilon) \quad \text{as } \varepsilon \to 0.$$

Corollary 5.3. The normalized queueing process for a system with bounded input

$$\zeta_N(t) = [\nu_N(t) - N\rho(t)]/\sqrt{N}.$$

Converges weakly as $N \to \infty$ to the Ornstein–Uhlenbeck diffusion process with the drift function

$$a(t, u) = -u(\lambda + \mu)$$

and variance

$$B(\rho(t)) = B^2 + (\rho_0 - p)(\mu - \lambda)e^{-(\lambda+\mu)t}$$

where $B^2 := 2\lambda\mu/(\lambda + \mu)$. In the particular case when the initial condition $\rho_0 = p = \lambda(\lambda + \mu)$, that is,

$$\nu(0) = Np + \zeta^0\sqrt{N} + o(\sqrt{N}),$$

the limiting diffusion process is stationary with the constant variance

$$B^2 = 2\lambda\mu/(\lambda + \mu).$$

Alternatively, the stationary regime is established with an exponential rate as $t \to \infty$.

The stationary distribution of the limit diffusion process with the generator

$$L_0^0 \varphi(u) = -ua\varphi'(u) + B^2\varphi''(u)$$

is a Gaussian distribution with the variance

$$\sigma^2 = B^2/a,$$

that is, the density of the stationary distribution is

$$\pi(u) = \frac{1}{\sigma\sqrt{2\pi}} e^{-u^2/2\sigma^2}.$$

For the system with bounded input

$$\sigma^2 = 2\lambda\mu/(\lambda+\mu)^2.$$

5.5.4 Repairman problem

An extensive literature is devoted to this problem [6, 9]. There are n identical working devices which operate independently, m spare devices and r repairing facilities. If one of n operating devices fails, it is sent to the repair facility and is replaced by one of the spare devices. There are at most n devices operating at any time.

The working time of every device has exponential distribution with intensity λ. If r, or more, devices are being repaired, the failed device has to join a queue at the repair facility and wait for service. Let $\nu_n(t)$ be a number of devices undergoing or waiting for repair at time t.

The diffusion approximation of the queueing process $\nu_n(t)$ in the repairman problem essentially depends on correlation between parameters n, m, and r.

There are considered three different variants.

Corollary 5.4. [6]. Let the following condition be fulfilled:

$$m = nm_0, \quad r = nr_0, \quad m_0 > r_0 \qquad (5.80)$$

where m_0 and r_0 are fixed numbers and, in addition,

$$r_0\mu < \lambda. \tag{5.81}$$

Then the normalized queueing process

$$\zeta_n(t) = [\nu_n(t) - n\rho]/\sqrt{n} \tag{5.82}$$

with the centering constant

$$\rho = 1 + m_0 - r_0\mu/\lambda$$

converges weakly as $n \to \infty$ to the diffusion process $\zeta^0(t)$ with the drift function $a(u) = -u\lambda$ and variance $B = 2r_0\mu$.

Proof. The normalized queueing process can be represented as follows:

$$\zeta_n(t) = \sqrt{n}(\kappa_n(t) - \rho)$$

where $\kappa_n(t) := \nu_n(t)/n$. This scaled process satisfies the inequality

$$0 \leq \kappa_n(t) \leq 1 + m_0.$$

The intensities of jumps of the queueing process $\nu_n(t)$ are different on the intervals

$$[0, r), \quad [r, m), \quad [m, n + m].$$

They are

$$\lambda_n(k) = \begin{cases} n\lambda, & k \leq m, \\ (n + m - k)\lambda, & k \geq m, \end{cases} \tag{5.83}$$

and

$$\mu_n(k) = \begin{cases} k\mu, & k \leq r, \\ r\mu, & k \geq r. \end{cases} \tag{5.84}$$

According to the basic assumption (5.65) of Theorem 5.4 the intensity functions are transformed to ($\varepsilon^2 k = v$, $\varepsilon = 1/\sqrt{n}$)

$$\lambda_\varepsilon(k) = \lambda(v) = \begin{cases} \lambda, & v \leq m_0, \\ (1 + m_0 - v)\lambda, & v \geq m_0, \end{cases}$$

and

$$\mu_\varepsilon(k) = \mu(v) = \begin{cases} v\mu, & v \le r_0, \\ r_0\mu, & v \ge r_0. \end{cases}$$

By assumption (5.80) and (5.81) of Corollary 5.4 and taking into account the condition

$$\nu_n(t) = k, \quad \zeta_n(t) = u$$

the normalized correlation has the form

$$\varepsilon^2 k = v = \rho + \varepsilon u.$$

Therefore, the intensity functions are considered on the intervals

$$\lambda(v) = \begin{cases} \lambda, & \varepsilon u \le -(1 - r_0\mu/\lambda), \\ (1 + m_0 - v)\lambda, & \varepsilon u \ge -(1 - r_0\mu/\lambda), \end{cases}$$

and

$$\mu(v) = \begin{cases} v\mu, & \varepsilon u \le -(1 - r_0\mu/\lambda) - (m_0 - r_0), \\ r_0\mu, & \varepsilon u \ge -(1 - r_0\mu/\lambda) - (m_0 - r_0). \end{cases}$$

The condition (5.80) of Corollary 5.4 gives consideration to the intensity functions on the interval

$$\varepsilon u \ge -(1 - r_0\mu/\lambda),$$

that is

$$\lambda(v) = (1 + m_0 - v)\lambda, \quad \mu(v) = r_0\mu.$$

Now, by Theorem 5.4, we calculate the parameters of the limiting diffusion.

$$\begin{aligned} C(v) &= \lambda(v) - \mu(v) \\ &= (1 + m_0 - r_0\mu/\lambda - v)\lambda \\ &= (\rho - v)\lambda, \end{aligned}$$

the drift function

$$a(u) = uC'(\rho) = -u\lambda,$$

and the variance

$$B = B(\rho) = \lambda(\rho) + \mu(\rho) = 2r_0\mu.$$

A more complicated situation arises under another asymptotical assumption on correlation between the parameters m, r, and n.

Corollary 5.5. [9] Let the number of spare devices m and the number of repairing facilities r satisfy the following asymptotical representation as $n \to \infty$:

$$m = n\rho + m_0\sqrt{n}, \quad r = n\rho + r_0\sqrt{n} \qquad (5.85)$$

where

$$\rho = \lambda/\mu < 1, \quad m_0 < r_0$$

are fixed.

Then the normalized queueing process

$$\zeta_n(t) = [\nu_n(t) - n\rho]/\sqrt{n}$$

converges weakly as $n \to \infty$ to the diffusion process with the drift function

$$a(u) = \begin{cases} -u\mu, & u \leq m_0, \\ -u(\lambda + \mu) + m_0\lambda, & m_0 \leq u \leq r_0, \\ -u\lambda + m_0\lambda - r_0\mu, & u \geq r_0, \end{cases} \qquad (5.86)$$

and the variance $B = 2\lambda$.

Proof. We start with the same representation (5.83)–(5.84) of the intensities of jumps of the normalized process $\nu_n(t)$, given by (5.82). But now, according to assumption (5.85) the intensity functions are transformed to the following form

$$\lambda_\varepsilon(v) = \begin{cases} \lambda, & u \leq m_0, \\ (1 + \rho + \varepsilon m_0 - v)\lambda, & u \geq m_0, \end{cases}$$

and

$$\mu_\varepsilon(v) = \begin{cases} v\mu, & u \leq r_0, \\ (\rho + \varepsilon r_0)\mu, & u \geq r_0. \end{cases}$$

Remind that, according to normalized condition, (5.82), that is

$$v = \rho + \varepsilon u.$$

According to the basic assumption (5.65) of Theorem 5.4 we obtain

$$\lambda_\varepsilon(v) = \lambda(v) + \varepsilon\lambda_1(v)$$

where

$$\lambda(v) = \begin{cases} \lambda, \\ (1 + \rho - v)\lambda, \end{cases} \qquad \lambda_1(v) = \begin{cases} 0, & u \leq m_0, \\ m_0\lambda, & u > m_0, \end{cases}$$

and
$$\mu_\varepsilon(v) = \mu(v) + \varepsilon\mu_1(v)$$

where
$$\mu(v) = \begin{cases} v\mu, \\ \rho\mu, \end{cases} \qquad \mu_1(v) = \begin{cases} 0, & u \le r_0, \\ r_0\mu, & u > r_0. \end{cases}$$

By (5.68) we get the centering velocity

$$\begin{aligned} C(v) &= \lambda(v) - \mu(v) \\ &= \begin{cases} \lambda - v\mu, & u \le m_0, \\ \lambda - v\mu + (\rho - v)\lambda, & m_0 \le u \le r_0, \\ \lambda - \rho\mu + (\rho - v)\lambda, & u \ge r_0, \end{cases} \end{aligned}$$

and

$$\begin{aligned} b^1(v) &= \lambda_1(v) - \mu_1(v) \\ &= \begin{cases} 0, & u \le m_0, \\ m_0\lambda, & m_0 < u \le r_0, \\ m_0\lambda - \rho\mu, & u > r_0. \end{cases} \end{aligned}$$

Hence,

$$C'(v) = \begin{cases} -\mu, & u \le m_0, \\ -(\mu + \lambda), & m_0 \le u \le r_0, \\ -\lambda, & u \ge r_0. \end{cases}$$

Applying Theorem 5.4 by (5.71) we obtain that the drift function is

$$a(u) = uC'(\rho) + b^1(\rho).$$

Hence, we get representation (5.86). In conclusion, it is easy to calculate the variance
$$B = B(\rho) = \lambda(\rho) + \mu(\rho) = 2\lambda.$$

There exists another asymptotic correlation between the parameters in the repairing problem.

Corollary 5.6. [9] Let the following asymptotic relations take place:

$$m = m_0\sqrt{n}, \quad r = np + r_0\sqrt{n}, \quad \text{as } n \to \infty, \tag{5.87}$$

with $p = \lambda/(\lambda + \mu)$ and fixed values of numbers m_0 and r_0.

The normalized queueing process

$$\zeta_n(t) = [\nu_n(t) - np]/\sqrt{n}$$

converges weakly as $n \to \infty$ to the diffusion process with the drift function

$$a(u) = \begin{cases} -u(\lambda + \mu) + \lambda m_0, & u \le r_0, \\ -u\lambda + \lambda m_0 - \mu r_0, & u \ge r_0, \end{cases} \qquad (5.88)$$

and the variance $B = 2\lambda\mu/(\lambda + \mu)$.

Proof. The intensity functions of jumps of the queueing process $\nu_n(t)$ are the same as in Corollary 5.4 and 5.5

$$\lambda_n(k) = \begin{cases} n\lambda, & k \le m, \\ (n + m - k)\lambda, & k \ge m, \end{cases}$$

and

$$\mu_n(k) = \begin{cases} k\mu, & k \le r, \\ r\mu, & k \ge r. \end{cases}$$

Taking into account the normalized condition $(u = \zeta(t))$

$$\nu_n(t) = np + u\sqrt{n}$$

and the asymptotical assumption (5.87), the intensity function $\lambda_n(k)$ can be transformed as follows:

$$\lambda_\varepsilon(v) = \begin{cases} \lambda, & u \le m_0 - p\sqrt{n}, \\ (1 + \varepsilon m_0 - v)\lambda, & u > m_0 - p\sqrt{n}. \end{cases}$$

Therefore, it is sufficient to consider the intensity function $\lambda_\varepsilon(u)$ only on the interval $u > m_0 - p\sqrt{n} \to -\infty$ as $n \to \infty$:

$$\lambda_\varepsilon(v) = \lambda(v) + \varepsilon\lambda_1(v)$$

where

$$\lambda(v) = (1 - v)\lambda, \qquad \lambda_1(v) = m_0\lambda.$$

The intensity function $\mu_n(k)$ can be transformed by assumption (5.87) as follows

$$\mu_\varepsilon(v) = \begin{cases} v\mu, & u \le r_0, \\ (p + \varepsilon r_0)\mu, & u \ge r_0. \end{cases}$$

Hence, we obtain

$$\mu_\varepsilon(v) = \mu(v) + \varepsilon\mu_1(v)$$

where

$$\mu(v) = \begin{cases} v\mu, \\ p\mu, \end{cases} \qquad \mu_1(u) = \begin{cases} 0, & u \le r_0, \\ r_0\mu, & u > r_0. \end{cases}$$

Now, applying the algorithm of diffusion approximation taken from Theorem 5.4 we calculate

$$\begin{aligned} C(v) &= \lambda(v) - \mu(v) \\ &= \begin{cases} (1-v)\lambda - v\mu = \lambda - v(\lambda + \mu), & u \le r_0, \\ (1-v)\lambda - p\mu = 1 - p\mu - v\lambda, & u > r_0. \end{cases} \end{aligned}$$

and

$$\begin{aligned} b^1(v) &= \lambda_1(v) - \mu_1(v) \\ &= \begin{cases} m_0\lambda, & u \le r_0 \\ m_0\lambda - r_0\mu, & u > r_0. \end{cases} \end{aligned}$$

The equilibrium point p is defined by the relation (5.73)

$$C(p) = (1-p)\lambda - p\mu = 0.$$

That is, $p = \lambda/(\lambda + \mu)$. According to Remark 5.6 the drift function is the following

$$a(u) = ua + a_1$$

where

$$a = C'(p) = \begin{cases} -(\lambda + \mu), & u \le r_0, \\ -\lambda, & u > r_0, \end{cases}$$

$$a_1 = b^1(p) = \begin{cases} m_0\lambda, & u \le r_0, \\ m_0\lambda - r_0\mu, & u > r_0, \end{cases}$$

and the variance

$$
\begin{aligned}
B &= B(p) = \lambda(p) + \mu(p) \\
&= (1-p)\lambda + p\mu \\
&= 2\lambda\mu/(\lambda+\mu).
\end{aligned}
$$

der

Remark 5.7. The queueing process in the repairman problem has various diffusion approximations under different assumptions of asymptotical relations between parameters m, r and n. In a view of this circumstance, the problem arises how to choose asymptotical representation for the initial parameters to get an optimal rate of convergence.

5.6 Markovian queueing networks

5.6.1 Collective limit theorems in R^N

A Markovian network is determined by a finite number N of nodes. In every node there is a Markov queueing system. The customers are transferred from one node to another according to a marching matrix

$$
P = [p_{kr}; \ 1 \le k, \ r \le N]
$$

where p_{kr} is the probability of the transfer customer from the k-th node to r-th node.

By the values of queueing process

$$
\nu(t) = (\nu_k(t); \ 1 \le k \le N)
$$

it is meant that there are a fixed number $\nu_k(t)$ of customers in the k-th node at time t. So the queueing process $\nu(t)$ takes values in the Euclidean space R^N of the dimension N.

Consider the open network that means there exists at least one k for which

$$
p_{k0} := 1 - \sum_{r=1}^{N} p_{kr} > 0.
$$

In other words, the customer is leaving the k-th node with probability p_{k0}. The functioning of the network in time is determined by the intensity vector-functions $\lambda(u) = (\lambda_k(u); \ 1 \leq k \leq N)$ and $\mu(u) = (\mu_k(u); \ 1 \leq k \leq N)$ where $u = (u_k : \ 1 \leq k \leq N) \in R^N$. The customers arrive from the outside at the k-th node with the intensity of exponentially distributed time $\lambda_k(u)$, and the exponentially distributed time of service in the k-th node has intensity $\mu_k(u), \ 1 \leq k \leq N$. Without loss of generality assume that the virtual passages aren't possible: $p_{kk} = 0$ for all k. Begin diffusion approximation of the queueing process of the network by introduction of a small series parameter $\varepsilon > 0$. The main assumption is that the intensity functions depend on the series parameter ε in such a way:

$$\lambda_\varepsilon(k) = \lambda(\varepsilon^2 k),$$

$$\mu_\varepsilon(k) = \mu(\varepsilon^2 k)$$

where $k = (k_k; \ 1 \leq k \leq N)$ is a vector of values of the queueing process $\nu(t)$. The normalized queueing process is considered in a similar form as for a queueing systems

$$\zeta^\varepsilon(t) = \varepsilon \nu^\varepsilon(t/\varepsilon^2) - \varepsilon^{-1}\rho(t) \tag{5.89}$$

with some deterministic positive-valued vector function

$$\rho(t) = (\rho_k(t); \ 1 \leq k \leq N).$$

Use the following notation. A vector with the superscript "d" is a diagonal matrix, for example:

$$\mu^d := [\mu_k \delta_{kr}; \ 1 \leq k, r \leq N]$$

where $\delta_{kr} = 0$, if $k \neq r$ and $\delta_{kk} = 1$ are Kronecker symbols.

Theorem 5.5. (*Collective*) [1] Let the following conditions be valid:

a) the vector function $\lambda(u)$ and $\mu(u)$ are continuously differentiable with bounded first derivatives;

b) there exists the unique positive solution of the differential equation

$$d\rho(t)/dt = c(\rho(t)), \quad \rho(0) = \rho_0 \tag{5.90}$$

where

$$c(u) := \lambda(u) + \mu^*(u)[P - I]; \tag{5.91}$$

c) the marching matrix $P = [p_{kr}; \ 1 \leq k, \ r \leq N]$ is irreducible and invertible;

d) the initial values of the normalized queueing process weakly converge in probability: $\zeta^\varepsilon(0) \Rightarrow \zeta^0$ as $\varepsilon \to 0$.

Then the normalized queueing process converges weakly

$$\zeta^\varepsilon(t) = \varepsilon \nu^\varepsilon(t/\varepsilon^2) - \varepsilon^{-1}\rho(t) \Rightarrow \zeta^0(t) \quad \text{as} \quad \varepsilon \to 0$$

to the Ornstein–Uhlenbeck diffusion process $\zeta^0(t)$ with the generator

$$L_t^0\varphi(u) = u^*c'(\rho(t))\varphi'(u) + \frac{1}{2}\text{Tr } B(\rho(t))\varphi''(u) \tag{5.92}$$

The covariance matrix $B(u)$ is determined as follows:

$$B(u) = c^d(u) - [\mu^d(u)Q + Q^*\mu^d(u)], \tag{5.93}$$

where

$$Q := P - I.$$

Besides, $\zeta^0(0) = \zeta^0$.

Proof of Theorem 5.5. First of all it is necessary to calculate the intensities of transfers of the normalized queueing process $\zeta^\varepsilon(t)$. We introduce the notation

$$\delta_k := (\delta_{kr}; \ 1 \leq r \leq N)$$

where δ_{kr} is the Kronecker symbol. Note that the normalized form of the queueing process (5.89) determines the following relation between variables $k = \nu^\varepsilon(t/\varepsilon^2)$, $v = \rho(t)$, and $u = \zeta^\varepsilon(t)$:

$$\varepsilon^2 k = v + \varepsilon u.$$

Now it is possible to determine the jump intensities of the normalized Markov queueing process $\zeta^\varepsilon(t)$:

$$Q^\varepsilon(u, \ \delta_k) = \lambda_k(v + \varepsilon u)$$

$$Q^\varepsilon(u, \ -\delta_k) = p_{k0}\mu_k(v + \varepsilon u)$$

$$Q^\varepsilon(u, \ \delta_r - \delta_k) = p_{kr}\mu_k(v + \varepsilon x). \tag{5.94}$$

The first formula gives the intensity of increase by one of the k-th component

$\nu_k(t)$: a customer arrives at the k-th node of the network. The second formula gives the intensity of decrease by one of the k-th component $\nu_k(t)$: a customer leaves the network. The third formula gives the intensity of increase by one of the r-th component $\nu_r(t)$ and decrease by one of the k-th component: a customer transfers from the k-th node to the r-th node of network.

So the first two moments of jumps of the normalized queueing process (5.89) can be calculated. By formula (5.30) the first moment in the following form is obtained:

$$b^\varepsilon(v, \ u) = \sum_{k=1}^N \delta_k[\lambda_k(v + \varepsilon u) - p_{k0}\mu_k(v + \varepsilon u)]$$
$$+ \sum_{k,r=1}^N (\delta_r - \delta_k)p_{kr}\mu_k(v + \varepsilon u).$$

Transforming the second term,

$$\sum_{k,r=1}^N (\delta_r - \delta_k)p_{kr}\mu_k(v + \varepsilon u)$$

$$= \sum_{k=1}^N \delta_k \left[\sum_{r=1}^N p_{rk}\mu_r(v + \varepsilon u) - (1 - p_{k0})\mu_k(v + \varepsilon u) \right]$$

So the first moment has the following representation:

$$b^\varepsilon(v, \ u) = \sum_{k=1}^N \delta_k \left[\lambda_k(v + \varepsilon u) - \mu_k(\rho(t) + \varepsilon x) \right.$$
$$\left. + \sum_{r=1}^N p_{rk}\mu_r(v + \varepsilon u) \right].$$

or, using the notation (5.91), in the vector form:

$$b^\varepsilon(v, \ u) = c(v + \varepsilon u)$$

Using Taylor's formula, the asymptotic representation of the first moment is:

$$b^\varepsilon(v, \ u) = c(v) + \varepsilon u^* c^{'}(v) + \varepsilon\theta_\varepsilon(v, u)$$

where $\|\theta_\varepsilon(t,\ u)\| \to 0$ as $\varepsilon \to 0$.

Hence, according to the notation in (5.30),

$$b(v) = c(v), \quad b^1(v,\ u) = u^* c'(v).$$

By Theorem 5.2 it can be concluded that the centering function $\rho(t)$ is determined by the solution of the equation (5.90) and the drift function of the limiting diffusion process is $b(v,\ u) = u^* c'(v)$. The second moment of jumps of the normalized queueing process (5.89) with the intensities of jump (5.94) is calculated in a more complicated way. By formula (5.31)

$$B^\varepsilon(v,\ u) = \sum_{k,r=1}^{N} \delta_k \delta_r^* [\lambda_k(v + \varepsilon u) + p_{k0}\mu_k(v + \varepsilon u)] +$$

$$+ \sum_{k,r=1}^{N} (\delta_r - \delta_k)(\delta_r^* - \delta_k^*) p_{kr}\mu_k(v + \varepsilon u)$$

The second term is transformed as follows:

$$\sum_{k,r=1}^{N} (\delta_r - \delta_k)(\delta_r^* - \delta_k^*) p_{kr}\mu_k(v + \varepsilon u) =$$

$$= \sum_{k,r=1}^{N} \delta_k \delta_k^* \left[(1 - p_{k0})\mu_k(v + \varepsilon u) + \sum_{r=1}^{N} p_{rk}\mu_r(v + \varepsilon u) \right] -$$

$$- \sum_{k,r=1}^{N} \delta_r \delta_k^* [p_{kr}\mu_k(v + \varepsilon u) + p_{rk}\mu_r(v + \varepsilon u)]$$

So the form of the second moment is such:

$$B^\varepsilon(v,\ u) = \sum_{r=1}^{N} \delta_k \delta_k^* [\ \mu_k(v + \varepsilon u) + \lambda_k(v + \varepsilon u)$$

$$+ \sum_{r=1}^{N} p_{rk}\mu_r(v + \varepsilon u) \Big]$$

$$- \sum_{k,r=1}^{N} \delta_r \delta_k^* [p_{kr}\mu_k(v + \varepsilon u) + p_{rk}\mu_r(v + \varepsilon u)]$$

or, using notation (5.93), the second moment is:

$$B^\varepsilon(v,\ u) = B(v + \varepsilon u)$$

Applying the Taylor formula, it is seen that asymptotic representation of the second moment has the following expression:

$$B^\varepsilon(v,\ u) = B(v) + \theta_\varepsilon(v,\ u).$$

By Theorem 5.2, the covariance matrix of the limiting diffusion process is defined by relation (5.93). The proof of Theorem 5.5 is completed.

5.6.2 Markov queueing networks

Two types of Markov queueing networks are considered and the superposition of independent Markov processes on a finite-dimensional phase states space.

Network of $[M/M/\infty]^N$ type. The Markov queueing systems of type $M|M|\infty$ are considered in each of N nodes of network. Let the intensities of arrival times at the nodes of the network be determined by the vector $\lambda = (\lambda_k;\ 1 \le k \le N)$ and let the intensities of service times in nodes of the network be determined by the vector $\mu = (\mu_k;\ 1 \le k \le N)$.

Corollary 5.7. The normalized queueing process of network of type $[M|M|\infty]^N$,

$$\zeta^\varepsilon(t) = \varepsilon\nu^\varepsilon(t/\varepsilon^2) - \varepsilon^{-1}\rho(t)$$

with the centering function is determined by the solution of the equation

$$d\rho(t)/dt = c(\rho(t)) \quad \rho(0) = \rho_0, \tag{5.95}$$

where

$$c(u) := \lambda + u^*Q, \quad Q := \mu^d(P - I), \tag{5.96}$$

under the conditions c) and d) of Theorem 3 weakly converges as $\varepsilon \to 0$ to the Ornstein–Uhlenbeck diffusion process with the drift function

$$a(u) = u^*Q \tag{5.97}$$

and the covariance matrix

$$B(t) = c^d(\rho(t)) - [\rho^d(t)Q + Q^*\rho^d(t)] \tag{5.98}$$

Proof. For the network of type $[M|M|\infty]^N$ the intensity functions $\lambda(u)$ and $\mu(u)$ in Theorem 3 are determined as follows:

$$\lambda(u) = (\lambda_k;\ 1 \le k \le N) \quad \mu(u) = (u_k\mu_k;\ 1 \le k \le N)$$

So, by formulas (5.90) and (5.91), (5.95) and (5.96) are obtained and then (5.97) and (5.98), by formulas (5.92) and (5.93).

Network of $[M|M|1|\infty]^N$ type. The Markovian queueing systems of $M|M|1|\infty$

type with intensities of arrival times λ_k and service times μ_k, $1 \leq k \leq N$, are situated in the nodes of the network.

Corollary 5.8. The normalized queueing process of network of type $[M|M|1|\infty]^N$

$$\zeta^\varepsilon(t) = \varepsilon \nu^\varepsilon(t/\varepsilon^2) - \varepsilon^{-1}\rho(t)$$

with the centering function $\rho(t) = \rho_0 + Ct$, $C = \lambda + \mu^* Q$ where $Q := [P - I]$, weakly converges, as $\varepsilon \to 0$, to the diffusion process with zero mean and covariance matrix

$$B = C^d - \mu^d[Q + Q^*]. \tag{5.99}$$

Proof. By Theorem 5.5, using the intensity vector function

$$\lambda(u) = (\lambda_k;\ 1 \leq k \leq N), \quad \mu(u) = (\mu_k;\ 1 \leq k \leq N),$$

according to formulas (5.90) and (5.91) the equation for the centering function is

$$d\rho(t)/dt = C, \quad C = \lambda + Q.$$

Hence, the centering function is determined as $\rho(t) = \rho_0 + Ct$. By formulas (5.92) and (5.93), we can conclude that the drift function of the limiting diffusion process is identically equal to zero, and the covariance matrix is given by (5.99).

Superposition of the Markov process. There are n independent ergodic Markov processes $\kappa_i(t)$, $1 \leq i \leq n$ on the finite phase states space $E = \{0, 1, \cdots, N\}$ with the generator ma trix

$$Q = [q_{kr};\ 0 \leq k,\ r \leq N].$$

Introduce the counting process

$$\nu^{(n)}(t) = (\nu_k^{(n)}(t);\ 1 \leq k \leq N)$$

with components $\nu_k^{(n)}(t)$ which describe the number of Markov processes $\kappa_i(t)$, $1 \leq i \leq n$ in state k ($1 \leq k \leq N$). Let $\rho = (\rho_k;\ 0 \leq k \leq N)$ be the stationary distribution of Markov processes:

$$\rho Q = 0, \quad \sum_{k=1}^{N} \rho_k = 1.$$

Introduce the notations:

$$\rho^0 := (\rho_k;\ 1 \le k \le N), \quad q^0 := (q_{0k};\ 1 \le k \le N)$$

$$Q^0 := [q_{kr};\ 1 \le k,\ r \le N],$$

$$u^0 := \sum_{k=1}^{N} u_k.$$

Corollary 5.9. The normalized counting process

$$\zeta^{(n)}(t) = [\nu^{(n)}(t) - n\rho^0]/\sqrt{n} \tag{5.100}$$

weakly converges, as $n \to \infty$, to the Ornstein–Uhlenbeck diffusion process with the drift function

$$a(u) = u^*Q^0 - u^0q^0 \tag{5.101}$$

and the covariance matrix

$$B = -[Q^0 + Q^{0*}]. \tag{5.102}$$

Proof. The counting process (5.100) can be considered as the nor malized queueing process for the Markov network with marching matrix

$$P = [p_{kr};\ 1 \le k,\ r \le N],$$

$$p_{kr} := q_{kr}/q_k,$$

$$q_k := \sum_{r=0}^{N} q_{kr}.$$

Considering the normalized counting process (5.100) in such a form

$$\zeta^{(n)}(t) = \sqrt{n}[\kappa^{(n)}(t) - \rho^0],$$

where $\kappa^{(n)}(t) = \nu^{(n)}(t)/n$.

Note that $\sum_{k=0}^{N} \kappa_k^{(n)}(t) \equiv 1$ because

$$\sum_{k=0}^{N} \nu_k^{(n)}(t) = n.$$

Hence the following relation exists between two variables $u = \zeta^{(n)}(t)$ and $\kappa = \kappa^{(n)}(t)$:

$$u = \sqrt{n}(\kappa - \rho^0),$$

or

$$\kappa = \rho^0 + \varepsilon u, \quad \varepsilon := 1/\sqrt{n}.$$

Besides, $\sum_{k=0}^{N} u_k = 0$, i.e. $u_0 = -\sum_{k=1}^{N} u_k$. Now determine the intensities of jumps of the Markov process (5.100) under condition

$$\zeta^{(n)}(t) = u, \quad \text{or} \quad \kappa^{(n)}(t) = \kappa$$

in the following way:

$$Q^\varepsilon(u, +\delta_k) = q_{0k}(\rho_0 + \varepsilon u_0)$$

$$Q^\varepsilon(u, -\delta_k) = q_{k0}(\rho_k + \varepsilon u_k)$$

$$Q^\varepsilon(u, \delta_r - \delta_k) = q_{kr}(\rho_k + \varepsilon u_k). \tag{5.103}$$

Applying the pattern limit Theorem 5.5, calculate the first moment of jump

$$b^\varepsilon(u) = \sum_{k=1}^{N} \delta_k [Q^\varepsilon(u, +\delta_k) - Q^\varepsilon(u, -\delta_k)]$$

$$+ \sum_{k,r=1}^{N} (\delta_r - \delta_k) Q^\varepsilon(u, \delta_r - \delta_k)$$

$$= \sum_{k=1}^{N} \delta_k [q_{0k}(\rho_0 + \varepsilon u_0) - q_{k0}(\rho_k + \varepsilon u_k]$$

$$+ \sum_{k,r=1}^{N} (\delta_r - \delta_k) q_{kr}(\rho_k + \varepsilon u_k).$$

The second term is transformed in such a way:

$$\sum_{k,r=1}^{N} (\delta_r - \delta_k) q_{kr}(\rho_k + \varepsilon u_k)$$

$$= \sum_{k=1}^{N} \delta_k \left[\sum_{r=1}^{N} q_{rk}(\rho_r + \varepsilon u_r) - (q_k - q_{k0})(\rho_k + \varepsilon u_k) \right].$$

Therefore the first moment of jump is in the form

$$b^\varepsilon(u) = \sum_{k=1}^{N} \delta_k \left[q_{0k}(\rho_0 + \varepsilon u_0) + \sum_{r=1}^{N} q_{rk}(\rho_r + \varepsilon u_r) - q_k(\rho_k + \varepsilon u_k) \right].$$

Using the property of stationary distribution $\rho = (\rho_k; \ 0 \le k \le N)$:

$$q_k \rho_k = \sum_{r=0}^{N} \rho_r q_{rk}, \ a \le k \le N.$$

Transform the representation of the first moment of jump

$$b^\varepsilon(u) = \varepsilon \sum_{k=1}^{N} \delta_k \left(u_0 q_{0k} + \sum_{r=1}^{N} u_r q_{rk} - u_k q_k \right).$$

Or, in the vector form,

$$b^\varepsilon(u) = a(u)$$

where $a(u)$ is defined in (5.101). The second moment of jump of the Markov process (5.100) by formula (5.31) is calculated in the following way:

$$B^\varepsilon(u) = \sum_{k=1}^{N} \delta_k \delta_k^* [q_{0k}(\rho_0 + \varepsilon u_0) + q_{k0}(\rho_k + \varepsilon u_k)]$$

$$+ \sum_{k,r=1}^{N} (\delta_r - \delta_k)(\delta_r^* - \delta_k^*) q_{kr}(\rho_k + \varepsilon u_k)$$

The second term is transformed in such a way:

$$\sum_{k,r=1}^{N} (\delta_r - \delta_k)(\delta_r^* - \delta_k^*) q_{kr}(\rho_k + \varepsilon u_k)$$

$$= \sum_{k=1}^{N} \delta_k \delta_k^* \left[\sum_{r=1}^{N} q_{rk}(\rho_r + \varepsilon u_k) + (q_k - q_{k0})(\rho_k + \varepsilon u_k) \right]$$

$$- \sum_{k,r=1}^{N} \delta_r \delta_k^* [q_{kr}(\rho_k + \varepsilon u_k) + q_{rk}(\rho_r + \varepsilon u_r)].$$

Therefore the second moment of jump has the following asymptotic representation

$$B^\varepsilon(u) = - \sum_{k,r=1}^{N} \delta_r \delta_k^* [q_{kr} \rho_k + q_{rk} \rho_r] + \theta_\varepsilon(u).$$

It can be performed in the vector form

$$B^{\varepsilon}(u) = B + \theta_{\varepsilon}(u)$$

where B is defined in (5.102).

References

[1] **V.V. Anisimov and E.A. Lebedev.** (1992) *Stochastic queueing networks.* Kiev, Lybid. (in Russian)

[2] **I.I. Gikhman and A.V. Skorokhod.** (1973) *Theory of random processes.* Vol. 2. (in Russian)

[3] **I.I. Gikhman and A.V. Skorokhod.** (1982) *Stochastic differential equations and their applications.* Kiev, Naukova Dumka. (in Russian)

[4] **S.N. Ethier and T.G. Kurtz.** (1986) *Markov processes characterization and convergence.* John Wiley & Sons, Inc. . New-York–London–Sydney.

[5] **W. Feller.** (1966) *An Introduction to Probability Theory and Its Applications.* John Wiley & Sons, Inc. . New-York–London–Sydney.

[6] **D.L. Iglehart.** (1965) Limit diffusion approximation for many-server queue and the repairman problem. *J. Appl. Prob.*, **2**, 429–441.

[7] **J. Jacod and A.N. Shiryaev.** (1987) *Limit Theorems for Stoc hastic Processes.* Springer–Verlag.

[8] **J.N. Kovalenko, Yu.Yu. Kuznetsov, and V.M. Shurenkov.** (1983) *Random processes, Handbook.* Kiev, Naukova Dumka. (in Russian)

[9] **V.S. Korolyuk and L.V. Vavricovich.** (1988) Diffusion approximation of the Markov renewal reserving systems. *Cybernetics.* **5**, 97–100. (in Russian)

[10] **V.S. Korolyuk and V.V. Korolyuk.** (1994) Diffusion approximation of Markov queueing systems and networks. *Mathem. Research.* **72**, 209–230.

[11] **V.S. Korolyuk and A.V. Swishchuk.** (1995) *Semi-Markov random evolutions.* Kluwer Academic Publishers. The Netherlands.

[12] **V.S. Korolyuk and A.F. Turbin.** (1993) *Mathematical foundation of the state lumping of large systems.* Kluwer Academic Publishers. The Netherlands.

[13] **R.Sh. Liptzer and A.N. Shiryaev.** (1986) *Theory of Martingales.* Nauka. Moscow.

[14] **V.S. Korolyuk, A.V. Skorokhod etc.** (1981) *Manual de la teo ria de probabilidades y estadistica matematica.* Editorial Mir. Moscu. (in Spain)

[15] **E. Nummelin.** (1984) *General irreducible Markov chains and nonnegative operators.* Cambrige Univ. Press.

[16] **G. Papanicolaou, D. Strook, and S. Varadhan.** (1977) *Martingale approach to some limit theorems.* Durham, Duke Univ. Math. Ser. . Vol. 3.

[17] **A.N. Shiryaev.** (1980) *Probability.* Moscow, Nauka. (in Russian)

[18] **V.M. Shurenkov.** (1989) *Ergodic Markov processes.* Moscow, Nauka. (in Russian)

[19] **A.V. Skorokhod.** (1986) *Random processes with independent increments.* Moscow, Nauka. (in Russian)

[20] **A.V. Skorokhod.** (1989) *Asymptotic methods of the theory o f stochastic differential equations.* Providence. USA.

Index

Additive functional, 5, 89
algorithm
    average
        ergodic, 107
        splitting, 110
        double, 113
    diffusion approximation
        ergodic, 107
        splitting, 112
    phase merging, 43, 72
        with splitting, 75

Balance condition, 50

Cumulant, 90

Equation
    Chapman–Kolmogorov, 23
    evolutional, 102
    Markov renewal, 33
    renewal, 12
    semigroup, 61
evolution
    equation, 92
    random, 102
    parameter, 5

Filtration, 58
flow

Poisson, 12
    recurrent, 12
    superposition, 15
    thinning, 15

Generator, 94, 97

Markov
    chain, 23
        homogeneous, 23
        imbedded, 25
        uniformly ergodic, 37
    jump process, 26
    renewal equation, 33
    moment, 25
    property, 23
    system, 5
martingale, 58
    characterization, 61
    integrable
        square, 57
        uniformly, 57
    problem, 59
    square characteristic, 60, 62

Operator
    averaged, 107
    contracted, 51
    reducible–invertible, 43

perturbed, 50
perturbing, 50
potential, 46

Phase merging, 11
  ergodic, 66
  heuristic, 76
  principles, 78
  splitting, 71
phase space, 2
  contracted, 49
  measurable, 2
  splitting, 9, 40
process
  adaptive, 56
  alternating, 19
  auxiliary, 13, 34
  counting, 11
  defect, 13
  excess, 13
  point, 11, 34
  Poisson
    compound, 91
    pure, 91
  queueing, 141
  regenerative, 35
  renewal, 11
  Markov, 20, 24
  semi-Markov, 20
  shift, 92
  storage, 5, 99
perturbation problem, 50
  singularly, 50
projector, 46
property
  ergodic, 37
    uniformly, 37

Markov, 23
  regenerative, 35
potential, 46

Random
  evolution, 102
  factor, 29
  medium, 89
renewal
  equation, 12
  function, 12
  intensity, 12
  lifetime, 11
  limit theorem, 13, 38
  moment, 11, 25
  process, 11
  residual time, 27

Semigroup 60, 100
  equation, 61
semi-Markov
  kernel, 24
  matrix, 24
  process, 32
  system, 5
sojourn time, 5, 11, 25
  remaining, 13, 37
steady-state regime, 40
stochastic
  additive functional, viii, 89, 92
  flow, 11
  kernel, 24
  matrix, 24
stoppage
  intensity, 80
  probability, 78
  state, 78

    time, 80
stationary
    distribution, 38
    independent increments, 89
    renewal time, 14
    sojourn time, 21
system
    double renewal, 81
    dynamic, 5, 99
    merged, 10
    Markovian, 4
    protective, 86
    queueing, 141, 145
    semi-Markov, 5
    stochastic, 4
    supporting, 10
    two-component, 78

Weak convergence, 63, 108